Understanding Animal Husbandry

Understanding Animal Husbandry

Florence Carter

STATES
ACADEMIC PRESS
www.statesacademicpress.com

Understanding Animal Husbandry
Florence Carter
ISBN: 978-1-63989-534-2 (Hardback)

STATES
ACADEMIC PRESS

Published by States Academic Press,
109 South 5th Street,
Brooklyn, NY 11249, USA

Cataloging-in-Publication Data

Understanding animal husbandry / Florence Carter.
p. cm.
Includes bibliographical references and index.
ISBN 978-1-63989-534-2
1. Animal culture. 2. Livestock. 3. Domestic animals. I. Carter, Florence.
SF61 .U53 2022
636--dc23

Table of Contents

Preface

The branch of agriculture which is concerned with the animals raised for obtaining their meat, milk, fiber, eggs and other products is known as animal husbandry. Practices related to livestock management and selective breeding are essential parts of this discipline. A large number of industries such as the milk and dairy industry, honey manufacturing, silk farming, etc. are dependent on animal husbandry. According to the animal being raised, animal husbandry practices are classified into dairy farming, fish farming, poultry farming, and bee farming. Rearing of animals can be done both extensively and intensively. In extensive farming, animals are allowed to roam freely across the farm. Intensive farming practices involve the use of cages, barns, high density feedlots and climate controlled buildings for rearing of animals. This book unfolds the innovative aspects of animal husbandry which will be crucial for a holistic understanding of the subject matter. It is compiled in such a manner, that it will provide in-depth knowledge about the theory and practice of this field. This book will serve as a source of information to a broad spectrum of readers.

A detailed account of the significant topics covered in this book is provided below:

Chapter 1- The branch of science dealing with the controlled cultivation, management and production of farm animals for utility, sport and research is termed as animal husbandry. Livestock management is an interdisciplinary field which involves managing and supervising farm animals. It requires the knowledge of animal husbandry, animal science and business. This chapter serves as an introduction and a guide to the subject of animal husbandry.

Chapter 2- Animal nutrition involves feed preparation and focuses on the dietary nutrition needs of animals. Animal feed is the food given to domestic animals. It includes straw, hay, silage, oils and mixed rations, compressed and pelleted feeds, etc. The chapter closely examines the key concepts of animal nutrition, feeding and health to provide an extensive understanding of the subject.

Chapter 3- The management and breeding of domestic or farm animals for obtaining meat and other products is known as livestock farming. There are various branches of livestock farming such as cattle farming, sheep farming, goat farming, horse farming, pig farming, rabbit farming, poultry farming, etc. The topics elaborated in this chapter will help in gaining a better perspective about these branches of livestock farming.

Chapter 4- In animal breeding, animals with desirable genetic traits are selected and bred to maintain these traits in future generations. The process of producing offspring by mating of genetically related individuals is known as inbreeding. Outbreeding refers to the matings between individuals who have no common ancestors, or are distantly related. This chapter has been carefully written to provide an easy understanding of the varied facets of animal breeding.

Chapter 5- Any material derived from the body of an animal is termed as animal product. They are broadly classified as food and non-food products. A few of the products which are derived from animals are meat, milk, wool, fur, fiber, etc. This chapter has been carefully written to provide an extensive understanding of the different aspects of these animal products.

I would like to make a special mention of my publisher who considered me worthy of this opportunity and also supported me throughout the process. I would also like to thank the editing team at the back-end who extended their help whenever required.

Florence Carter

1

Introduction to Animal Husbandry

The branch of science dealing with the controlled cultivation, management and production of farm animals for utility, sport and research is termed as animal husbandry. Livestock management is an interdisciplinary field which involves managing and supervising farm animals. It requires the knowledge of animal husbandry, animal science and business. This chapter serves as an introduction and a guide to the subject of animal husbandry.

Animal husbandry is the science or a branch of agriculture that is concerned with the production, taking care of, and breeding of farm animals. Humans are dependent on both plants and animals for food; we raise animals for different food items including eggs, milk and meat, and other products like fiber, wool, leather, etc. We deal with a branch of agriculture to provide day-today care, selective breeding, and thus raising livestock (of a variety of domestic animals) and the process is known as animal husbandry.

It has a long history when animals were initially domesticated during the Neolithic revolution. Earlier, cattle, goats, sheep, and pigs were raised on farms on a low scale. After the 18th century, agriculturalists rapidly took animal husbandry to the next level and yielded more meat, wool, and milk day by day. A wide variety of species like horses, rabbits, guinea pod, and water buffalo were also used in some parts of the world. Aquaculture of molluscs, crustaceans, and fish was on the rise and insect farming was also adopted. Contemporary methods Co define animal husbandry depends on the type of land available and production systems adapted Co these. Intensive animal farming that involves maximum production with minimum cost has been adopted by the modern methods that make available thousands of chickens and high-density feedlots. This helps in increasing the yield per area of land with an investment of a good amount of money and labor.

Importance of Animal Husbandry

Animal rearing is important for humans as they attain a wide range of food products having high nutrient values. They meet the commercial requirements of high demand for food such as dairy needs from cows, goats, and buffaloes. They are called milch animals as they provide us milk which is full of protein and other vitamins and minerals.

The importance of animal husbandry further extends Co another set of animals that are important to humans for eggs and meat such as hens, ducks, goose, goat, fish, etc. They are also bred for meat which is rich in protein, iron, fats, vitamin B12 and Zinc. All of these nutrients are essential to enhance metabolic rate, promote fullness, and reduce hunger. Marine animals like fish are also believed to keep a person happier and healthier as these have Omega-3 fatty acids, DHA, and vitamin D.

Animal husbandry takes care of the management and breeding of animals and is a large scale business. A farm or a specific region is built for providing shelter, food, and undertaking culturing processes. It also involves apiculture i.e. bee agriculture and aquaculture i.e. marine animals culturing.

Types of Animal Husbandry

Depending on the types of farming, there are various types of animal husbandry explained as follows:

- Poultry Farming: It is a kind of animal husbandry that involves raising domesticated birds including chickens, turkeys, ducks, and geese for producing meat and eggs for food. Large farming is done for poultry in a disease-free environment to obtain healthy and nutritious food from them. Maintaining sanitation and hygiene is important during poultry farming as animals or birds engage in producing a lot of body waste materials. The feces are also used in producing manure or fertilizers for soil. Poultry farming enables massive employment to people that help in the improvement of their economy, especially farmers.
- Dairy Farming: It is a class of agriculture where long-term production of milk takes place that leads to sales after its processing. Various products like curd, yogurt, cheese, butter, and cream can be obtained through the management of dairy animals like sheep, goats, buffaloes, and cows. These dairy farming animals are called milch animals. Healthy animals are opted for dairy farming and are inspected regularly for their overall well-being, physically as well as mentally. Animals are milked by machines or by hands and preserved well to make it industrially applicable that can then be used commercially.
- Aquatic Farming (Pisciculture): Aquatic farming called Pisciculture or fish farming, it is also one of the major types of animal husbandry, which is involved in the rearing of fish and other marine creatures like prawns, in enclosures and controlled manner is called aquatic farming. It is raised in controlled water bodies like tanks, open nets, cages, and ponds. It enables disease-free rearing at a fast pace and makes available commercial supply. Extensive aquaculture and intensive aquaculture are the two types of fish farming known where the former is based on local photosynthetic production and the latter is based upon the external food supply for aquatic animals.

- Bee Farming (Apiculture): Apiculture refers to the bee farming practice. So, it involves taking care and management of honey bees to produce honey and the wax. Honey possesses wonderful healing properties and can reduce ulcers, regulate blood sugar, lessen digestive problems, treat coughs, and sore throats. Apiculture ensures a safe place for the bees to work and live. Bees are essential for us too as these are engaged in pollination for our food sources. Having bee farming in place, it provides an environment for studying their habitat and behavior.

Advantages of Animal Husbandry

The role of animal husband, in human welfare can be described as follows:

- Dairy products producing animals like cows, sheep, and goats are the major sources of milk and milk products including butter, cheese, yogurt, cottage cheese etc.
- Many animals including chicken, g.t duck, pigs, and buffaloes are used for their meat because of their highly nutritious value.
- Livestock is used to control the growth of weeds on agricultural land as the dry shrubs that are prone to fir are eaten up by various animals. This reduces the risk and harm that can be faced with unprecedented situations. So. It is important for land management.
- Wool and leather are the fibre by-products from animals like sheep and camel. When these animals are reared in a vast umber, the products can be obtained largely too.
- The importance of animal husbandry also lies in the fact that the excreta, bones, and blood of animals can be used as manure or fertilizers via composting. These can be spread on fields to enhance crop yield.
- Animals are helpful for ploughing fields. Military functions, and transporting goods from one place to another. Example: Horse, yak, and donkey.

Good Farming Practices

A number of common themes run through all levels of farm management and recur often in the principles elaborated below. They are:

Legal Obligations

Farmers should be aware of, and comply with, all legal obligations relevant to livestock production, e.g. disease reporting, record keeping, animal identification and carcass disposal.

Record Keeping

When a problem arises in an enterprise, be it a disease, a chemical hazard issue or a physical safety matter, record keeping is central to any effort to trace the source of the problem and eliminate it. Hence, as far as is practicable, farmers should keep records of:

- All animal populations on the farm (groups or individuals as relevant).
- All animal arrivals, including their identification marks or devices, origin and date of arrival, to ensure that movements of incoming animals are traceable to their source.
- Movements of animals around the enterprise.
- Changes to feeding or health regimes, and any other management changes that may occur.
- Origin and use of all feeds, drugs, disinfectants, herbicides and other consumable items used on the farm.
- Known diseases/infections, diseased/infected animals and mortalities, as far as possible giving details such as dates, diagnoses (where known), animals affected, treatments and results.

Animal Identification

Animal identification and the ability to trace animals have become important tools to ensure food safety and improve management. Identification of animals may be on an individual or group basis, and connections between properties as a result of animal movements should be able to be deduced from good record keeping and animal identification. Where a food safety incident occurs, it should be possible to determine the source of the problem and to take appropriate action. The ability to trace animals at least one step forward and one step back from the current holding is recommended.

Hygiene and Disease Prevention

Measures aimed at preserving cleanliness, preventing pathogen build-up and breaking possible pathways of transmission are essential in the management of any modern farming enterprise, regardless of the species or the farming system. While the use of animal manure, animal slurry and human sewage sludge for fertiliser purposes is becoming increasingly common, enabling higher crop yields as well as sensible waste management, it may facilitate the transmission of food-safety-related diseases within or between herds or directly to humans.

Therefore systems for animal or human waste usage for fertiliser purposes should take into consideration relevant treatment methods as well as specific holding times before animals are allowed onto treated pastures. Suggested holding times are directly related to climatic conditions in the region in question (e.g. die-off of pathogens is faster at higher temperatures). As a general rule, neither animal nor human waste should be used on plants intended for direct human consumption unless it has been appropriately treated. Precautions should aim at:

- Reducing contact between healthy animals and potentially infected animals.
- Maintaining the hygiene and safety of all facilities.
- Ensuring the health of all workers on the farm and the implementation of hygienic working procedures.
- Taking all appropriate measures to prevent contamination by vehicles entering and traversing the property.
- Minimising contact between livestock and professional or other visitors, and taking all hygienic measures necessary to reduce the possible introduction of pathogens and contaminants.
- Ensuring overall health of livestock through good nutrition and reducing stress.
- Maintaining an appropriate population density for the species and age group in question, either by following locally enforceable measures or by obtaining appropriate advice from recognised experts.
- Keeping records of animal populations in facilities/on farms.

Training

Husbandry measures and techniques are ever-changing. Competent Authorities are encouraged to assess training needs amongst stakeholders and promote necessary training. Farmers and farm managers should:

- Actively seek and use relevant training opportunities for themselves and their workers.
- Be aware of any training courses that may be compulsory in their countries and regions.
- Keep records of all training undergone.

Animal Health Management

Addressing Biohazards

As a general principle, closed farming systems and all-in all-out systems are recommended from a food safety and biosecurity point of view. Owners or managers of livestock should:

- Establish a working relationship with a veterinarian to ensure that animal health and welfare and disease notification issues are addressed.

- Seek veterinary assistance to immediately investigate any suspicion of serious disease.
- Comply with regulations concerning restrictions on animal movements.
- Separate diseased from healthy animals such that transmission of infection does not occur and, where necessary, cull diseased animals.
- Practice breeding and selection such that animals well suited to local conditions are raised and detailed breeding records are kept.
- Acquire animals (including breeding stock) only from sources with a known, safe health status, where possible with supporting health certificates from veterinarians.
- Source fresh or frozen semen, ova and embryos from sources with a known, safe health status, accredited by the Competent Authority of the country of origin, with appropriate health certification.
- Keep records of all breeding stock, semen or embryos used on the premises, the animals upon which they were used, the breeding dates and outcomes.
- Keep newly arrived animals separate from resident stock for an appropriate period to monitor them for diseases and infestations in order to prevent transmission of such conditions.
- Ensure that wherever necessary, newly arrived animals are given time to adapt to new feeding regimes, are not overcrowded, and that their health is regularly monitored.
- Ensure that equipment and instruments used in animal husbandry are suitably cleaned and disinfected between each use.
- Effectively remove or dispose of dead and fallen stock where possible so that other animals cannot come into contact with carcasses and that carcasses do not contaminate the pasture or drinking water, and keep records of all such disposals.

Addressing Physical Hazards

Owners or managers of livestock should apply animal welfare practices in accordance with regulatory requirements, and in particular:

- Ensure that people working with animals are properly experienced and trained for the tasks they should perform.
- Ensure that facilities and equipment are properly designed and maintained to prevent physical injury.
- Ensure that animals are handled and transported appropriately.

Veterinary Medicines and Biologicals

Owners or managers of livestock should:

- Be aware of and comply with restrictions on medicines or biologicals for use in livestock.
- Use veterinary medicines and biologicals strictly in accordance with the manufacturer's instructions or veterinary prescription.
- Use antimicrobials only in accordance with regulatory requirements and other veterinary and public health guidance.
- Keep detailed records of the origin and use of all medicines and biologicals, including batch numbers, dates of administration, doses, individuals or groups treated and withdrawal times. Treated individuals or groups should be clearly identified.
- Maintain required storage conditions for veterinary medicines and biologicals.
- Ensure that all treatments or procedures are carried out using instruments that are appropriate and correctly calibrated for the administration of veterinary medicines and biologicals. Dispose of used instruments (including needles) in a biosecure manner.
- Keep all treated animals on the farm until the relevant withdrawal times have expired (unless the animals should leave the farm for veterinary treatment) and ensure that products from these animals are not used for human consumption until the withdrawal periods have elapsed.
- Ensure that all handling or treatment facilities are safe and appropriate to the species in question, facilitate correct and calm handling and restraint, and that their construction is such that the likelihood of injury is minimised.

Animal Feeding and Watering

Owners or managers of livestock should:

- Acquire feed from suppliers who follow recognised good manufacturing practices.
- Manage the feed chain (transport, storage, and feeding) in such a way as to protect feed from contamination (biological, chemical, and physical hazards) and minimise deterioration. Feeds should be used as soon as possible and, if applicable, in accordance with label instructions.
- Ensure that only water of known and acceptable biological and mineralogical quality (i.e. fit for animal consumption) is used for watering stock.

- Keep records of all feeds and dates of acquisition and feeding; where possible the animals/groups of animals fed should be clearly recorded. Self-mixed feeds should have their ingredients and mixes recorded, as well as dates of feeding and animals fed.
- Where on-farm manufacture of feeds is practised, follow procedures designed to minimise contamination and prevent the inclusion of undesirable feed components. Where necessary, expert assistance should be sought.
- Ensure that nutritional levels are adequate to promote animal health, growth and production.
- Ensure that changes to feeding regimes are, where ever possible, gradual, and that the regimes are safe and follow acceptable feeding practices.
- Prevent animal access to places where feeds are stored and to places where hazardous chemicals are stored.

Addressing Biohazards

Owners or managers of livestock should:

- Ensure that antibiotics are not used in feed for growth promoting purposes in the absence of any public health safety assessment and recommendations.
- Ensure that ruminant protein is not fed to ruminants.
- Where appropriate, manage pastures by stocking rate and rotation to maintain healthy and productive livestock and reduce parasite burdens. Keep records of pasture rotation and other on-farm animal movements between pens, sheds etc.
- Regularly inspect and, when necessary, clean and disinfect feeding and watering facilities such as drinkers and troughs.
- Ensure that effluents are managed in such a way that drinking water sources are not contaminated.

Addressing Chemical Hazards

Owners or managers of livestock should:

- Use herbicides and pesticides judiciously and according to the manufacturer's instructions and applicable legislation such that animal exposure to these chemicals is minimised. Records of usage, including the date and location of application, should be kept.
- Ensure that when feed additives are used, that manufacturer's instructions as to dosage levels and withdrawal periods are followed, and that records of usage of such feed additives are kept.

Addressing Physical Hazards

Owners or managers of livestock should ensure that animals are not kept in sheds, pens or pastures where they are likely to ingest foreign objects and that all facilities are kept clean and free from metal objects, pieces of wire, plastic bags, etc.

Environment and Infrastructure

Owners or managers of livestock should:

- Ensure that where animals are confined, the housing or pens are constructed such that the basic needs of the animals are fulfilled especially with regard to ventilation, drainage, and manure removal. Walking surfaces should be level, non-slip, and all surfaces should ideally be washable.
- Locate farms in areas free from industrial and other pollution and sources of contamination and infection.

Addressing Biohazards

Owners or managers of livestock should:

- Ensure that farm layout and building construction provide for adequate separation of animals by production group as necessary.
- Ensure that buildings and perimeter fences are constructed so that contact with other livestock and wild animals is minimised.
- Maintain adequate separation between clean and contaminated materials (e.g. feed and manure).
- Ensure that systems that use animal or human waste for fertiliser purposes take into consideration relevant treatment methods as well as specific holding times before animals are allowed onto treated pastures.
- Ensure that effluent is properly disposed of and that facilities where animals are kept are an appropriate distance from any disposal points.
- Ensure that any bedding or litter is regularly renewed and that used bedding or litter is disposed of safely.
- Apply appropriate pest and vermin control measures, which may include the use of barriers such as nets or fencing, or the use of pest/vermin population control measures.

Addressing Chemical Hazards

Owners or managers of livestock should:

- Use chemical disinfectants and cleansers strictly in accordance with the manufacturer's instructions, ensuring that disinfected or cleaned surfaces and facilities are properly rinsed if necessary.
- Seek professional advice with regard to the use of disinfectants or cleansers.

Addressing Physical Hazards

Owners or managers of livestock should manage pastures such that livestock are not exposed to dangerous and impassable areas.

Animal and Product Handling

Addressing Biohazards

Owners or managers of livestock should:

- Ensure that all animals destined for slaughter are clean, healthy and fit to travel and have not had recent contact with diseased stock or infectious material.
- Apply short duration feeding regimes aimed at reducing the shedding of harmful bacteria by animals destined for slaughter.
- Ensure that contamination of animal products from animal and environmental sources during primary production and storage are minimised.
- Ensure that storage conditions maintain the quality of the products.
- Keep records of animals and animal products leaving the farm as well as their destination and date of dispatch.

Addressing Chemical Hazards

Owners or managers of livestock should:

- Ensure full compliance with existing legislation such that applicable maximum residue levels are not exceeded.
- Ensure that no animal destined for slaughter has been subjected to treatment for which the withdrawal period has not elapsed.

Addressing Physical Hazards

Owners or managers of livestock should:

- Ensure that mustering or catching and handling of animals prior to loading is carried out in a safe and humane manner.

- Ensure that loading facilities are appropriately constructed.
- Take the necessary care during animal loading so as to minimise injury.
- Handle products in such a way as to prevent damage.

Organic Animal Husbandry

Breeding animals which are strong and robust and adapted to the local conditions is a strategy which is important both for health promotion and disease prevention. Traditional breeds of farm animals may be a good starting point for organic animal breeding. Animals can be improved by selection of individuals specially adapted to their natural organic conditions. They can be crossbred with suitable new breeds, thus achieving animals with the desired positive aspects of both breeds.

For breeding, organic farming uses natural reproduction techniques, keeping bulls at the farm for natural mating. Although artificial insemination is allowed, we do not promote AI Techniques but rather breeding systems based on the integrity of animals, which AI techniques do not respect. Breeding in organic animal agriculture is based on the following requirements:

- Breeding systems shall be based on breeds that can reproduce successfully under natural conditions without human involvement.
- Artificial insemination is permitted.
- Embryo transfer techniques and cloning are prohibited.
- Hormones are prohibited to induce ovulation and birth unless applied to individual animals for medical reasons and under veterinary supervision.

Breeding Goals

Over the last decades, traditional breeds have been replaced by high performing ones in many regions. Similar to high yielding plant varieties, these new animal breeds usually depend on a rich diet (concentrates), a huge amount of water and optimal/protected living conditions, because they can be very susceptible to disease. It can be very expensive to keep such animals. Some of the conventional animals are even bred from lines which grew up on a huge amount of veterinary treatments, which also contributes to less robust animals and dependency on veterinary inputs. When focusing only on one breeding goal such as high yield, we often lose other abilities, which have to do more with robustness. Many organic farmers do not keep animals only for production, but also many other purposes and all of them should be considered in the breeding goals. Organic animal breeding should optimise the overall use of farm animals, with considerations given to the local conditions and available fodder: breeding goals for poultry and cattle breeding.

Breeding Goals

The «ideal» organic poultry breed

- Feeding on kitchen wastes and farm by-products
- Satisfying egg production
- Useful as meat
- Good health, good resistanc against diseases

The «ideal» organic cattle breed

- Utilising roughage and farm by-products
- Satisfying milk production
- High fertility
- Good resistance against diseases
- Long life with continuous production

Life Production a Relevant Breeding Goal in Organic Herds

When comparing the production of different breeds of cows, the daily or yearly production or daily weight gain are used. However, high performing so-called 'exotic' breeds usually have a shorter life span than traditional breeds with lower production. The life time milk production of a cow which e.g. gives 8 litres per day over 10 years therefore would be greater than the one of a high yielding cow which yields 16 litres per day, but is culled or dies after 4 years. It costs money and risks raising an animal from new born or newly hatched to a productive animal, and once it is 'stabilized' as adult, it is evident that it fits well into the particular environment. Finally, having an animal for more years - e.g. a cow for 10 lactations - require a very good framework and care for the animal, and having this as a goal reflect such goals of giving the animals sufficient living conditions for living a long life. This is relevant to include in the breeding goals.

Table: Comparing the economic performance of two different breeds.

Template for comparing the economic performance of two cattle breeds.		
Item	Breed A	Breed B
Investment: Costs of a calf Costs of a cow		
Maintenance: Fodder purchases per year Veterinary costs per year		

Milk production: Litres per day Litres per year Productive years in life Litres per life		
Other uses: Meat Draught Dung		

Keeping Animals Outdoor and Indoor

Housing

It is crucial to keep animals in ways which allow them to move freely. In particular, the operator shall ensure the following animal welfare conditions:

- Sufficient free movement and opportunity to express normal patterns of behavior, such as space to stand naturally, lie down easily, turn around, groom themselves and assume all natural postures and movements such as stretching, perching and wing flapping.
- Sufficient fresh air, water, feed and natural daylight to satisfy the needs of the animals.
- Access to resting areas, shelter and protection from sunlight, temperature, rain, mud and wind adequate to reduce animal stress.

This applies both on indoor or outdoor keeping, and it is emphasized that this is also possible when tethering on grass. Specifically about outdoor keeping, 'All animals shall have unrestricted and daily access to pasture or a soil-based open-air exercise area or run, with vegetation, whenever the physiological condition of the animal, the weather and the state of the ground permit. Such areas may be partially covered. Animals may temporarily be kept indoors because of inclement weather, health condition, reproduction, specific handling requirements or at night. Lactation shall not be considered a valid condition for keeping animals indoor'.

In many smallholder tropical farming systems, this is difficult, and there are many ways of trying to solve this. On holdings where, due to their geographical location and structural constraints, it is not possible to allow free movement of animals, tethering of animals may be allowed for a limited period of the year or of the day. In such cases, animals may not be able to turn around freely.

The pros and cons for shed feeding versus grazing is in relation to feeding; this is an area where the integration of animals into the farm should be very carefully thought through, including how the different elements can support each other.

Herd animals such as cows which always live in a group, must not be kept in isolation from other animals of same species, although acknowledging that this will not be possible in some smallholder farms where they e.g. only have one animal. In special cases, isolation is acceptable, such as when keeping males from females or in case of disease. If there is only one animal of a species on a farm, good human contact or contact with other animal species can help a lot. Outdoor access as well as indoor housing conditions will be managed differently for different animal species and this will be specifically dealt with under each of the animal species.

Indoor Housing Conditions

With the exception of nomadic lifestyles, most farm animals are temporarily kept in sheds. The combination of animal husbandry and other farm activities requires that the animals are kept under control to avoid damage to crops. As emphasized above, they should be allowed free movements. The 'building construction provides for insulation, heating, cooling and ventilation of the building, ensuring that air circulation, dust levels, temperature, relative air humidity, and gas concentrations are within levels that are not harmful to the livestock'. This includes the requirement that there should be sufficient aeration, but no draught. The Norms also say that 'no animals shall be kept in closed cages', and 'animals are protected from predation by wild and feral animals'. Lastly, it is also emphasized that the amount of artificial light should be limited - this applies more for large industrialized farms than for most smallholders.

Hygiene must be emphasized: There should be as little contact as possible between the animals and their own manure, both for their own health and wellbeing, and for human health reasons. Construction material and production equipment that might significantly harm human or animal health shall not be used. Generally, the construction or equipment should not harm at all! For economic reasons, sheds can be built with simple, locally available materials. Many countries have a rich tradition of shed constructions, and have developed the most efficient and appropriate shed systems for the conditions of the region. If techniques of this heritage are combined with the above principles, a locally adapted and at the same time animal friendly system may be obtained.

Bedding Material

Bedding materials should be used in sheds to keep the lying area of the animals soft, dry, and clean. This is important for their health and wellbeing. The beddings need to be replaced from time to time, and can be used in the gardens as compost, which recycles nutrients to the plant and add improves the soil (which e.g. enhance its water retention capacity). Beddings can be of straw, leaves, twigs, husks or other locally available material. They can be replaced daily or kept for several months while adding fresh material on top, and letting the material in the 'mat' compost. (The mat is the mattress, if you apply material on top of other material without removing the old straw or leaves or whatever material you have before applying new. Then a 'mat' (e.g. a straw mat) or a 'mattress' is built up).

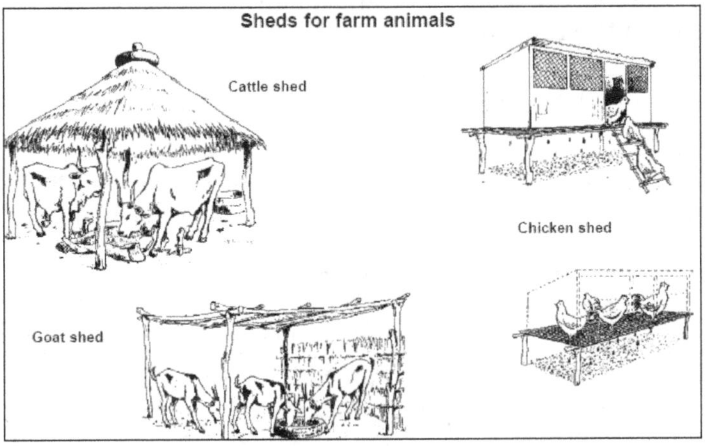

Traditional simple sheds in Senegal: Cattle shed, goat shed, chicken shed.

Feeding Animals

Organic animals receive their nutritional needs from organic forage and feed of good quality. Animals shall be offered a balanced diet that provides all of the nutritional needs of the animals in a form allowing them to exhibit their natural feeding and digestive behaviour.

Organic animal feeding should be mainly based on fodder produced on the farm, and this is an important aspect of the harmonic and well-balanced organic farming systems at least 50% of the feed must be from the farm or from other organic farmers in the area. The health and wellbeing of animals is strongly linked to the feeding. Fodder production should be integrated into the crops and other plant productions on the farm.

Organic animals must be fed organic feed - that means the feed is without pesticides and mineral fertilizer and as a part of a farming system which farm in accordance with the organic principles (not organically certified). Exceptions can be made if there is a very limited amount of organic feed in the area. From the farm itself, it is permitted to use feed produced on the farm as 'organic feed' to the farm's own animals during the first year of organic management.

Support the Health of Animals with a Well-balanced Diet

It is important for all animals to have a well-balanced diet which meet their needs. A part from physiological maintenance, the dietary requirements of the animal will also depend on whether the animal is kept for milk, beef or draught power. Seen from an economical perspective, it will always make most sense to limit the number of animals and give them all sufficient feed, if limited resources are available. Likewise, different animals require different amounts and composition of feed in different life situations.

Minimum Space and Water Consumption

Where livestock are housed, the minimum "on-ground" density for the indoor area shall be not more than the following:

Type of animal	Minimum space per animal	Water consumption in litres/day per animal
Bovine and Equine (adults) Cattle/Donkey/ Horse	4.0 m²/animal	30-80, donkeys and mules need twice a day 10-25 or as much as they can drink.
Camels	-	Daily 15-30 or every 5-8 days as much they can drink: up to 100 litres or 1/3 of bodyweight.
Sheep and goats (adults)	1.5 m²/animal	5-20
Porcine (pigs) (>40 kg) Sow with piglets	1.0 m²/animal 3.0 m²/animal	15-25 10-25
Poultry (adults) chicken/ duck/geese/turkey	6.0 m²/animal	0.5-0.75
Rabbits	0.3 m²/animal	50-150 Millilitre per kg bodyweight.

Water should be available at all times (except for camels, they can do with water every 5-8 days) and be clean and fresh. Remember that young animals also need water. Even when they are milk fed, it is not always fulfilling their needs for liquids, especially not if active and if it is warm or hot and dry, or maybe even windy.

Innovative Techniques in Animal Husbandry

The increased world population is demanding more reliable quality livestock products the number of farms is decreasing but the number of animals for per farm and animal production is increasing In addition to this trend livestock production problems also increasing. The solution of these problems comes from multidisciplinary studies from very different fields such as technology. In large enterprises it is not possible to obtain the expected performance without using technology and automation systems from animals with very high genetic values. Daily work on livestock farming is simple in and standard application routinely Data monitoring in the modern dairy farm enables the ongoing control of production, animal health, and welfare. However, as the number of animals increases, error burden and work load increase. Successful livestock farmers will be capable of rapidly adapting their infrastructures to exploit changes in technology for better production. Mechanism and automation systems offer options in front of the user in intense competition for convenience.

Currently, most data is extracted manually, yet manual observation is gradually being replaced by many milking systems by automated recording (milk yield, milk conductivity, activity recording and body weight measurements) leading to better data, both in quantity and quality. The number of farms automation systems has increased rapidly since 1980. Almost any medium- to large-sized farmers can benefit from enhanced automation. There are many opportunities for facilities in automation technologies and systems. Today livestock farmers increasingly use robots on production or algorithms to optimize their farm management decisions. Technological developments are creating a new automation system in which smarter and more flexible work possibilities in livestock production. The automation of animal husbandry and integration of on-farm systems and processes have a key role to play in facilitating the process of meeting each of important challenges for competitive market. The main technology is electronic recording, milking, heat detection auto-weighing, auto-drafting, genetic improvement, feeding, barn optimization, and health monitoring, livestock housing and equipment designs. These technologies provide to dairyman many opportunities to make easier and more convenient their decisions about dairy future plans.

Current Technology Applications

The benefits of new technology are plentiful and include increased cost efficiency, improved animal welfare, improved working conditions, better production monitoring (e.g. remote monitoring, access to real-time data) and improved provision of important production data. The new technology means producers can work easier and improve cattle welfare, production efficiency, and profitability. Technologic developments provide more efficient, profitable and fast solutions for farmers to get on time process using management and direct manipulation possibilities. Continuous monitoring of disease and its careful management is essential for the well-being of an animal management. This can be achieved through the detection of early stages and, subsequently, the detection and treatment of the infection. Automation today is super-sophisticated technology and software as well as complicated machinery. A number of computer-assisted image analysis applications are being developed for more convenient animal husbandry. The latest computer programs can identify and classify sounds of animal for specific situations. These applications could be used to monitor the welfare of animals and provide early identification of disease, physiologic status, and abnormality.

The main technology that livestock farmer's requirements met is electronic records, milking, heat detection walk-over-weighing, auto-drafting, genetic improvement, feeding, barn environment optimization, and health recording etc. Some sensors are currently available for this purpose, but they do not fulfill all demands. Also, with advances in proteomics and genomics, new biomarkers are being discovered, allowing the disease to be detected at earlier stages. This will lead to assays with higher sensitivity, which can provide additional quantitative information on the level of inflammation 'on-site' and 'on-line' and which is also faster and less expensive. These technologies provide to dairyman many opportunities to make easier and more convenient their decisions about dairy future plans.

Breeding and Genetics

In dairy farms which very high genetic value of breeding animals cannot get the expected performance without the use of latest technology? Dairy cattle herd management programs if can be used as effectively, dairy farming will have many advantages for consumer, farmer and also animals. Genetic information and type evaluation of herd members and bulls are particularly suitable for expanded electronic updating. However, to obtain these advantages from this system required to have knowledge of the functions and effective use of the functions. The large amount of data in the obtained on many issues related to animals, herd management, and an individual unless used in decisions about animals, ensuring the heavy data flow, record keeping or assessment will not give the expected results.

Breeds in animal husbandry have changed a lot with the use of breeding and gene technology. Till 1980s livestock products demands have been met by breed substitution, cross-breeding, and within-breed selection. But these demands in future are to be met using new techniques such as such as artificial insemination and more specific selection techniques. Genomic selection provides more possibilities for the higher rate of genetic gain in the livestock sector. After all genomic breeding values will be calculated from the genetic marker, rather than from pedigree and phenotypic information in near future. The genome maps for poultry and cattle is completed and these developments provide new opportunities for animal breeding and animal models. Those DNA-based tests for genes or markers affecting traits that are difficult to measure currently, such as meat quality and disease resistance, will be particularly useful. But genetic resources still important for helping livestock adapt to changing the climate.

Native breeds are to genetic insurance against future challenges. In combination with modem reproductive technologies, there is potential to use frozen and stored germplasm (genetic resource banks) to support conservation measures for the maintenance of genetic diversity in threatened species. Besides the direct application of technologically advanced reproductive procedures, modern approaches to non-invasive endocrine monitoring play an important role in optimizing the success of natural breeding programs. A separate progeny-test category may be developed for farms that collect all data electronically and have those data monitored closely. Automated data collection along with parentage verification offers substantial opportunities for genetic improvement of overall economic merit. Nowadays biological samples are sent laboratory for genetic analysis to identify the relevant genes responsible for productive parameters. Also, selective breeding can reduce the need for alternative methods.

Computer and Internet Usage

New technology in computers, biotechnology and scientific discoveries regarding ruminant nutrition and genetics provide the basis for accelerated progress in milk production for those dairy farmers that adopt these technologies. 10 years ago most dairy farmers focused their attention solely on animal husbandry practices. The use of computers for farm management in dairy sector started in as early in 1990s in many developing countries. As personal computer was developed and the price has dramatically declined, more and more farmers began to use computers by themselves in the last decade. But generally, computers have been used by producers with larger farms. Small-scale farmers bypassed the technology because of its cost and their lack of knowledge about computer use in farming.

Many computer programs were described, by which data on data in dairy herds may be processed. The some computer software is designed for timely and direct convenience to farmers. Thus, the breeder can evaluate the monthly lots of data using many formulas with high accuracy using this software. It can also be programmed for annual report for detail evaluation of herd evaluation. In addition to all these, daily milk yields feed consumption, pregnancy check; inseminated cow list can be programmed for daily work routine. In recent years there is a form of high interest to cattle breeding and this is leading to the establishment of intensive farms.

The only criteria for the life cycle continuity of these intensive farms would be on maximum profitability and competitiveness ability on market. This concept mainly related to forceful usage of knowledge, technology and management at intensive farms and small enterprises and cattle breeding organizations. Whenever the farmers meet any problem in order to refer to an organization for learning new solutions and the absolute result most probably they prefer to share with farmers who are more experienced for them. But developed countries heavily use computer and internet that is the main way to reach information. Meanwhile in undeveloped or developing countries, several reasons limit using computer and internet these are listed as high financial cost, difficulties to use technology, loss of knowledge to economic benefits, hesitate to use new technologies, lack of education, strict personality, poor infrastructure, lack of personal experience and not enough time to spent. On the other hand, the country wide effect of the communication instruments extends to 80% and this is enough to eliminate most of the reasons which are mentioned above. If the farmer evaluates the benefits of using computer and internet they will replace this technology in farm management.

Electronic Identification

The Electronic identification system is started 1970s. However, current laws deal with the visual, readable markings that are placed on the animal. There are numerous animal ID technologies available to livestock producers. Radio frequency identification (RFID) will likely be used to identify cattle. These devices have an electronic number that will be unique for an individual animal and link that animal to the database. Electronic ear tags, injectable transponders and boluses with a transponder, inside in the reticulum are the latest technology for animal identification technology.

Many types of RFID tags (boluses, ear tags, injectable glass tags) are used subcutaneous placement for animal identification. These systems work using radio frequency for sending data. Boluses retain in the first two stomachs of the ruminants and accepted as safe for animal health. They can be administered even to lambs after weaning at the fifth week and the retention rate can reach 100%. The injectable transponders, on the other hand, can be applied easily after birth, while the preferable locations differ in each animal species. These technologies (implants, ear tags, and rumen boluses) are available on the market for cattle farmers. All these devices have special chip system for sending data for the base computer for evaluation. These devices have some specific components on their system regarding storing and evaluating data used for evaluating herd data. Some electronic tags has reader which can be receive and store the required many data for evaluation. Some of tag works transferring the number to another storage system for another evaluation stage. Data sends using antenna for transfer data on the system. From a technological point of view, RFID tags can be grouped in two categories according to the carrier frequency band: LF (low frequency) tags function at 125–134.2 kHz, whereas HF (high frequency) tags function at 13.56 MHz. Electronic scales may be justified as a way to determine body condition score automatically.

Another technology which is very useful for farmers is electronic weighing system. An easy and powerful electronic weighing system that accurately measures cattle weight. So farmers can monitor cattle performance easily and continuously. These system established on the road the watered or cattle squeeze. Stored information sends to the main computer for evaluation. Complimenting this is auto-drafting, where cattle going through a race are automatically separated on the basis on age, sex, or weight, or any other criteria the producer preferences.

Milking Automation

Milking automation system is also involving the dairy sector at 1990s. Suitable objective measuring systems are needed in animal husbandry to quickly and safely recognize illness, normal estrus cycle, quiet heat or stress in animals. An automatic milking system requires a completely different management system for milking, feeding, cow traffic, cow behavior and grazing, but also for safeguarding milk quality and animal health. Electronic devices or sensors are the tools that need to take over the human visual inspection for abnormality. In order to develop sensors to detect abnormal milk a definition of abnormal milk is still basic requirements. Sensors have been in the market for a long time, but their use in milking systems is quite new. Because milks were being evaluated by milkers during milking. However, with the development of intelligent milking systems, the use of sensors in the milking systems has become widespread.

The milking robots equipped with sensors to detect signs of mastitis which measures the many characters of the abnormal milk pH, Somatic cell count, milk acidity, milk conductivity etc. systems also can be regarded milking specifications of the system such as parlor performances, milking efficiency etc. Simple automatic cup removal devices monitor the milk flow rate from individual cows and at a threshold, the milking vacuum is shut off and the system is activated to withdraw the cups from the cow. Post-milking teat disinfection is an established component of many mastitis control strategies. This is normally performed manually in many farmers using either a pressure operated spray lance or more a dip cup. Behavior meter also installed to the milking systems for animal monitoring. The behavior meter continuously records the lying time, lying bouts and the activity of the individual animals. The cow-behavior observations enable animal welfare assessment in different environmental conditions and stressful situations, as well as reproductive and health status. Other options to separation gate usage at automatic management systems.

The cattle separation is a risky and challenging activity that needs to be done frequently. If milkers also make an animal separation, the milking efficiency and parlor performances decrease. Reducing the need and risk of this workforce for separation is an important advantage. The grouping and separation of cattle in the big herd constitutes an enormous workload for the farmers. Electronic separation gates are not common in many cattle farms.

Removing the labor required to separate animals can have a significant impact on the performance of the handling and management operations. To a lesser extent, diseased cows need to be brought to the attention of the dairy farmer. Some sensors are currently available for this purpose, but they do not fulfill all demands. When an operator is involved with animal separation, other tasks are not being done and performance suffers. With larger herds, identification and drafting of individuals are major tasks. Automatic drafting is not routinely installed on many dairy farms. Electronic tongue technology gives more advantage for farmers for many aspects Electronic tongue used potentiometric chemical sensors. An array comprised sensors with plasticized PVC membranes with cross-sensitivity to inorganic and organic cations and anions, chalcogenide glass sensors, chloride-, potassium- and sodium-selective electrodes, and glass pH electrode. Automatic milking systems using newly developed sensors (NIR, SCC and LDH etc.) provide much faster and more effective results. Many biosensor search studies for mastitis diagnosis continue.

The current SCC and alternative methods for detection of mastitis. There are a lot of sensors which are used for good quality milk productions. Faster results have been achieved with the use of microchip technologies. In addition, with these technologies, you are ready to diagnose more successful mastitis with more effective tests and results with wider angle, more accurate results. All these each tests provide rapid mastitis detection. Milk conductivity and appearance of milk is used commonly on the farms. But other methods give early mastitis detection for the fast and accurate decision for cure disease.

A number of other methods using visible and other light spectra have shown promise in detecting milk abnormalities and measuring various components of milk. Table summarized the technology of main sensors used for mastitis detection.

Table: Research results of sensors technology used for mastitis detection.

Near infrared (NIR)	SCC in raw milk
Electro biochemical sensor using a screen-printed carbon electrode (SPCE)	Detect NAGase via its ability to convert the substrate 1-naphthyl N-acetyl-b-D-glucosaminidinase to 1-naphthol.
A gas-sensor array system, or 'electronic nose'	Interact with volatile substances, including sulfides, ketones, amines and acid.
To automatically determine the SCC based on measuring the DNA content of somatic cells	The DNA and histone levels can then be measured and correlated to the SCC.
PicoGreen	The DNA from somatic cells was incubated with PicoGreen, and the resulting fluorescence was measured using an optical sensor.
Competitive biosensor assay	Surface plasmon resonance to monitor the interaction between Hp, which was immobilized onto the chip surface, and hemoglobin (Hb).
Fluorescence was measured using an optical sensor	A chip for simultaneously monitoring pathogens, somatic cells and pH in raw milk samples.
Chemical-array-based sensor "electronic tongue'	To detect chloride, potassium and sodium ions released during mastitis in addition to inorganic and organic cations and anions.
Disposable microchips	The milk sample is mixed with a lysis solution to burst the somatic cells, and a fluorescent dye is added to stain the DNA.
Disposable device	On counting milk leukocytes.
Detection of the patterns of volatile metabolites produced	To identify different pathogens, such as *S. aureus*, coagulase negative staphylococci, streptococci and *E. coli*, and to determine infection-free udder quarters.
A lactate screen printed sensor	Elevated levels of lactate.
Biochips	Sensor-based platforms with the development of novel biomarkers could thus allow the diagnosis of the pre-clinical stage of mastitis.

Microfluidic CD-based assay device	After centrifugation on a conventional CD-player, the SCC can be measured based on the height of the cell pellet formed.
A biochip	Incorporated DNA amplification of genes that are specific for seven known mastitis-causing pathogens.
Microfluidic device	Integrates solid-phase extraction and NASBA has recently been reported for the identification of low numbers of *E. coli*.

The single and combined measures of 29,033 milkings to detect clinical mastitis and concluded that early warning is not reliable with sensors and software currently on the market. As of 2000 there were not yet sufficiently effective methods available to monitor characteristics of milk automatically so as to divert milk from unhealthy cows. Many farmers were still reluctant to rely on electronic devices to monitor cow health status. Automatic milking systems give much information about milk production, milking speed, milk acidity; milk conductivity etc. new sensor added some other new component such as milk progesterone level, milk temperature etc. But radio-frequency identification provides more possibilities for improving the reliability of collecting data.

Feeding Automation

Computer programmer designed many software for make best option for farmer to ration preparation. Optimal feeding programs can be done for advanced options such as live weight, racing, lactation period and animal feed stock information. These programs use data from the National Research Council in animal feed and feed content. Various systems for automated animal feeding will be used in many big dairy farms to get better production. They will comprise complete systems include each stage of feeding, feed preparation, mixing equipment and the installations for distributing feed. Feed components such as grass and maize/corn silage as well as mineral feed and feed concentrate will be loaded, mixed and delivered to the feed table built up there by the systems.

The Automation systems as simple consists of a control panel, a programmable command manager, a scale, a communication interface and finally all the needed equipment to organize the feeding process and feed provision to the animal of each age groups. Computer-controlled calf feeders have many advantages over traditional calf feeding methods. Calves carry a transponder, and it is possible to follow the daily intake of individual calves. Calves learn to use the computer-controlled milk feeding system fairly easily and this technology offers a significant reduction in labor cost (73%). These systems can be combined with automatic weighing and health observation system for calf welfare. Calves reared in a group-pen had fewer days of medication than calves in hutches fed milk-replacer from buckets twice a day. Electronic Concentrate Feeding system ensures that each cow is supplied with the exact ration of feed at the exact right time. The Belt Feeder feed distributor is the ideal introduction to the concept of automatic feed supply systems. Small, flexible, economical – the combination of a conveyor belt and sliding scraper. Grothmann et al. reported that the various technical approaches to automation. These are reported that the stationary systems such as conveyor belts and mobile systems such as self-propelled or rail guided feeder wagons. In addition to feeding system automation approaches, rumen activity sensors are very popular innovative techniques for cattle farmers to reduce metabolic disorders. When the sensitive cows exhibit increasing acidosis, this allows a farmer to adjust feeding to prevent major problems.

Many electronic sensors can be used for rumen pH and rumen temperature of cattle. Especially rumen bolus can work 100 days continuously and data stored every 15 minutes for future evaluation. The rumination activity is a good indicator of cattle health condition. A certain level of wellbeing is a prerequisite for rumination excitement and stress, states of anxiety and various diseases inhibit rumination. Another sensor used for collecting data for cow jaw movement to estimate chewing activity. This sensor works on the principle that the changing pressure of the animal is not detected during opening and closing of the mouth.

Health Observation

The big hazard for animal production is to disease outbreak. The disease can spread quickly in the confined conditions. Many diseases have specific signals for detection, animals to look for signs of stress, disease, and damage caused by many agents. They alert staff or, potentially, other systems to find the affected animals and identify them report to manger before the problem spreads. An animal disease has serious economic implications on farm productivity. Public institutions and private groups are working collectively to assist individuals in addressing society's stake in disease prevention and control. The right time detects disease three to 5 days' sooner, reduce treatment costs, reduce mortality rates, and improve production efficiency. The production, product quality, product composition, body condition, and behavior provide a good indication for the health status of animals. By closely monitoring normal pattern changes, the farmers ensure animal health status.

Many firms provided programs developed and provided by data collection and analysis products for monitoring animal behavior for the best early detection system. To monitor the health conditions of each cow the sensors are mounted on the cow. Sensor networks consist of several tiny, low price devices and are logically self-organizing ad hoc systems. The role of the sensor network is monitoring the health parameters of animals, gather and convey the information to other sink nodes. Sensors that collect data such as temperature, pH, etc., receive a lot of data, so it is possible to transmit data at intervals. Many new sensor technologies that will be useful in animal health and behavior are developed.

Another sensor usage results of an experiment in which a temperature sensor built into a bolus were placed in the rumen of a cow. On-farm scoring of behavioral indicators of animal welfare is challenging but the increasing availability of low cost technology now makes automated monitoring of animal behavior feasible. Furthermore, behavioral measures, such as the occurrence of aggression or stereotypic behavior, are important indicators of welfare problems. Including behavioral-based welfare criteria is, therefore, essential for an overall welfare assessment.

Reproductive Performances

Estrus detection technology; Average calving interval in cattle farm is the best criteria for comparisons for reproductive performances of the farms which is varying between 13 and 18 months, heat detection efficiency vary between 30 and 50% in most dairy herds. Research results showed that the 5–30% of the cows were not in or near oestrus when inseminated. Results of oestrus detection varied depending on the many factor such as threshold value, cow number, barn style, and the statistical method for data analysis. The detection error rates between 17 and 55% and indicate a large number of false warnings. As a result of satisfying oestrus detection and conception rates, purchase and maintenance costs of the oestrus detection system should charge off. A number of both inexpensive to expensive aids and technologies are available to meet some but not all of these criteria.

Traditionally, oestrus detection is performed by visual observation of the dairy herd in many countries but this procedure particularly difficult on large dairy farms because of short observation periods during feeding and milking. Galiç et al. reported that the effect of herd size on milk yield, calving age, lactation number, and calving interval is significantly important ($P < 0.01$) and small farms are generally more successful than large farms. Mean duration of oestrus was calculated as 13.5 h with a standard deviation of 2.3 h. As a result of technical progress in monitoring cows using computers, automatic oestrus detection has become possible. In many studies, different traits have been analyzed for utilization in automatic oestrus detection. The electronic systems are an electronic device that detects cows that stand to be mounted by a herd mate and provides a continuous monitoring of activity, radio telemetry is a computerized estrus detection devices. Also patches give another possibilities using mounting activity of cows. I a cow mount another cow then the transmitter is depressed and a signal sent to a receiver. During this time, date, time and duration of the mount stored and send to the main computer. On computer all these data evaluated and prepared for final decision.

Although costs associated with computerized estrous detection are higher than other methods, the benefits may pay off with increased estrous detection accuracy. Estrus detection errors can result huge economic loses for dairy farms. The economic loses vary \$2–\$6/day for dairy farms. But missing 1 cycle cost \$42 to \$126 for a cow. Using detection aids provide advantages because of the prevention of these losses. Pedometers are used to detect the estrus by storing past physical activity the current physical activity and comparing it previous activity data. After analyzing data programs prepare report for cow which is activity accepted as estrus. Beeper or flashing light is also using for alerts the farmer for control this cows.

Pedometers also used for estrus detection attached to the leg of the cow to measure the amount of her activity over a unit time span. Many pedometric systems are commercially available in the market. Also standing activity systems is commercially available in the markets. Standing activity activated by the mounting cow. Radio signal picked up by receiver and relayed to a buffer and a personal computer to analyzing of data. This system record cows number, standing time, date and duration to evaluation on time.

Voice identification processing can be used to detect estrus both economically (simple microphone) and accurately (over 94% accuracy), either as a stand-alone solution. One aid is a pressure sensitive device mounted on the back of each cow, which can be triggered when the cow stands for mounting. Pressure sensitive device is programmed when a certain number of valid mounts have been recorded a light give signals. The second one is effective aids for detecting standing estrus is a marker or teaser animal. Marker animals are worn marking device. When an animal in standing estrus is mounted by the marker animal, the chin-ball marker will rub against the animal in standing estrus, leaving marks on her back and rump. Mounting and standing activity are effective methods for estrus detection. There are many other methods available on the system such as cervical mucus, vaginal characteristics, temperature, blood flow, and hormone changes in blood and milk. But these methods not applicable on the farm level. Milk progesterone level is o good criteria for stage of the cycle or pregnancy. So it can be used for diagnose problem cows in herd. The behavior meter continuously records the animal behavior for many purposes (lying time, lying bouts and the activity of the individual cows). The cow-behavior enables animal welfare assessment in different environmental conditions and stress situations, as well as reproductive and health status.

Pregnancy Check

Pregnancy diagnosis is one of the most important factors to get ideal calving interval. The most common methods are rectal and transrectal ultrasonography of the reproductive tract. Both procedures are required training and time. An experienced practitioner using ultrasound can reliably diagnose pregnancy from 30 days gestation whilst an experienced veterinary is able to diagnose pregnancy from 35 days. Enzyme-linked immunosorbent assay (ELISA), radioimmunoassay (RIA) or latex agglutination (LA) tests use either blood or milk to detect a marker of pregnancy. Estrone sulfate, progesterone and glycoproteins are used for indicators of pregnancy in cattle. Estrone sulfate is a conjugated steroid product of estrone, is produced by the fetus and as such offer high specificity. The negative part of this test is to high rate of false negatives and the inability of the test to reliably diagnose pregnancies before 100 days of gestation, progesterone. Wireless system was designed to measure many characteristics of cows is also developed to detect early stage of pregnancy in multiple cows.

Barn Environment Control

Animal production starts at environment which is cow lived in. Many factors affect the sensitivity of cows to their surrounding environmental conditions. Latest technologies involve the use of sensors to collect data, followed by data analyses with the objective of enhancing the understanding of the system interactions, and developing control systems. Latest technologies aim to provide adequate data for producers and farmers to optimize the efficiency of their agricultural system, thus increasing the overall performance of the animals. There are many sensors for use at dairy barn environment control automation. Temperature and relative humidity sensors; airspeed sensors, carbon dioxide sensors, ammonia sensors and light sensors etc. When ambient temperature gets warmer than 25 °C cow begins use their energies to cool themselves down rather than to produce milk. The effects of heat stress on dairy cattle physiology and productivity have been well established. Milk yield can decrease by about 10 percent. At the same time, if the environmental factors for example air qualities are poor, milk production and quality can be affected adversely.

However high producing dairy cows need an optimal indoor climate throughout the year, to maintain high production levels. Barn environment is also important for the farm worker. While the thermoneutral zone for cattle ranges from −5 to 25 °C, the thermoneutral zone for people is shifted to higher air temperature ranges. Modern technology also helps to control barn environment which is many sensor installations to measure factors such as temperature, humidity, solar radiation, and luminosity over a large cultivated surface. These sensor and automation systems planned as a capable of recording and adapting to environmental conditions inside the barn. The variety of sensors monitors a wide-ranging range of parameters of interest. Automation systems not only can automate for temperature, but also have wind and rain sensors. The wind sensors feed wind speed data into the controller, which then adjusts curtain height to compensate for higher air transfer rates. The rain sensor can be programmed to close the curtain to a predetermined height when it rains to keep moisture off cows and stalls. Cows like bright environments. For this reason, equal illumination in barn improves milk yield. This is especially important during short winter days. For this reason the right kind of illumination planning, dimensioned to the size of barn, orientation of barn and roof material is very important for good illumination in barn.

Lighting is the most obvious change with the shift to automatize barn. Digitally controlled LEDs can extend the day, supplementing sun in autumn and winter. LEDs use less energy than traditional lamps, making artificial lighting economical. The availability of specialist barn luminaires makes it possible to tune the color. New technology provided is a self-regulating, micro-climate controlled environment for optimal animal growth and production. New technologic tools can monitor nearly every aspect of animal barn indoor environment. Incorporating the environment-sensing capability of wireless sensor networks into mobile monitoring systems can provide convenient control of the barn microclimate anywhere, anytime for more productive animal production.

Environmental sensors and other control facilities of the barn is first component of the barn automation. Secondly computerize system for monitoring and controlling for barn environment. And thirdly supports the communication between these two components.

Livestock Management

Livestock management involves managing farm animals and supervising farm workers. Livestock management requires knowledge of animal science and animal husbandry, as well as good business sense. The development and application of modern technology have enhanced the health of livestock or farm animals considerably. Some of the benefits include - high profits and well-organized livestock management. Moreover, the growing use of precise, robust and consistent livestock production tools has improved the monitoring, administration and movement of cattle, which has resulted in momentous productivity gains across the farming business.

Provide Nutritional Diet

Always select feedstuffs that are loaded with nutrients so that your cattle reach their best levels of production. A smaller amount of red meat and more feed variability will maximize the production of both - milk and meat. Feeding animals with less human food and letting them graze pastures that have rich-fibre content will improve their overall nutrition ingestion and resistance to unbearable diseases.

Give Right Food Supplements

Supplements have proved to improve animal health as well as productivity by encouraging the growth of helpful microbes in the rumen. Red clover is famous for an enzyme called Trifolium pratense that increases the capability of your livestock to take in dietary protein. The presence of clover in the feed helps in improving milk production and also increases the appetite of livestock. Tar brush supplements avoid gastrointestinal acidosis in livestock and reduce the release of greenhouses gasses. Azolla

Caroliniana (a water fern grown in ponds) offers added protein to animals that are lacking in protein. By including supplements to your farm animals daily feed is the best way to increase productivity of ruminant animals.

Use Technology

Technology facilitates reliable and accurate examining of animal health by providing appropriate and accurate interpretations in the form of structural figures. The arrival of imaging tools and livestock scales permit remote monitoring of animal health and help farmers in making informed decisions regarding rearing and feeding patterns. You will get several makes and models of livestock scales on the market depending on your resources. These automated cattle scales are long-lasting, cost-effective and accurate for:

- Checking animal health.
- Knowing their exact weight before breeding.
- Evaluating the conversion of feed.
- Assessing their performance.

Track Animal Performance

If you track the performance of livestock, you can easily recognize healthy breeds that do better and pick the unproductive breeds and use them for other things. Checking animal weight facilitates early recognition of diseases and prevents dangerous disease from spreading to other animals in the farm. Monitoring livestock will also help in selecting right weaning time and choosing animals for cross breeding.

Take Suitable Precautions

Animals that are imported from moderate climates often lack resistance to heat, humidity, ticks, parasites and tropical diseases hence it is essential to keep them in dirt free stalls where they stay away from disease vectors. Instead of letting your farm animals to graze, you must cut the silage and give them at the stalls. You can also purchase imported feed that assures quality.

Customize Practices Favorable to Local Climate

A number of people depend on livestock for livelihood but the benefits of rearing cattle could stop if the customary farming practices and conventional grazing are restored with industrial systems that do not take natural factors into deliberation. Altering farm practices for profits will work only when the local breeds, resources as well as feedstuffs are utilized and wastage is reduced. Local breeds when given a proper diet along with fresh water and supplements will stay healthy and be productive and profitable for the farmers.

Livestock Management Software

The livestock management software is a framework that helps the farmers to maintain and track their livestock right from when they are born till the time they are sold. It records the details of all the events that an animal goes through by saving all the details. With this system, it becomes easy to keep track of the number of farm animals, their performance and cost metrics and their financial tracking until sales.

Working of Livestock Management Software

Livestock farming can be extremely tiring, more so in case, it involves managing a wide range of animals at the farm. The farmers who use the livestock management software for their farming business get empowered because this system works to have better access to keeping up with the market trends while remaining at the top.

- Pasture Management: The Livestock management system assists in effectively moving the animals from one location to another while keeping a close track on their location. At a given point, you can have access to the information that tells the past and present location of the animals. This system is also helpful to make necessary breeding decisions because it has an inbred detection feature. It works to make sure that all livestock are raised in a healthy environment for ensuring the best farming practices.
- Elaborate Livestock Records: With the help of the livestock production system, it becomes possible to maintain and record essential information that includes breeding details, weight, gender, etc. All in all, all the little events that take place during the lifetime of all the livestock are recorded in the system. This helps in referring to the information whenever the point the owner wants.
- Feed Management: The software works to help the managers to maintain a consistent feeding plan while recording the amount of feed that is consumed by every single animal on the farm. This is an important feature which is of great importance in the feed conversion ratio. Calculating the cost of feed which is

being consumed from every single farm becomes possible with the assistance of this feature. Therefore, taking informed decisions that are based on real data becomes possible.

- Tracking the Dates: Regulated livestock farming that is practiced with the help of a livestock management software helps in tracking due dates for every single animal. Tracking the estimated calving period contingent on the time when the animals were exposed becomes possible. This is vital information that helps in monitoring the breeding practices that are carried out at the farm to maintain all animals with ease.
- Event Tracking: The various events in livestock farming such as animal sale, treatments, breeding events, births, and deaths can be tracked with the help of livestock management software. The software allows recording this important information so as to later compile the series of operations and events carried out on the farm.

It is a thing of the past when the managers used to stroll around with record book in their hands to take information that used to concern these animals being raised. In case you own a business of livestock farming, at that point, you have no choice however to get rid of the customary technique for maintaining a farm while putting resources into appropriate livestock management software. In addition to making things easier in the business, this software is the best way to make sure maximum productivity, profit, and great efficiency.

Impact of Livestock on Climate and Mitigation Strategies

The livestock sector requires a significant amount of natural resources and is responsible for about 14.5% of total anthropogenic greenhouse gas emissions. Mitigation strategies aimed at reducing emissions of this sector are needed to limit the environmental burden from food production while ensuring a sufficient supply of food for a growing world population.

Impact of Livestock on Climate Change

The most important greenhouse gases from animal agriculture are methane and nitrous oxide. Methane, mainly produced by enteric fermentation and manure storage, is a gas which has an effect on global warming 28 times higher than carbon dioxide. Nitrous oxide, arising from manure storage and the use of organic/inorganic fertilizers, is a molecule with a global warming potential 265 times higher than carbon dioxide. The carbon dioxide equivalent is a standard unit used to account for the global warming potential. Figure shows in carbon dioxide equivalents the greenhouse gas incidences that enteric fermentation and manure storage have across the main livestock species raised worldwide.

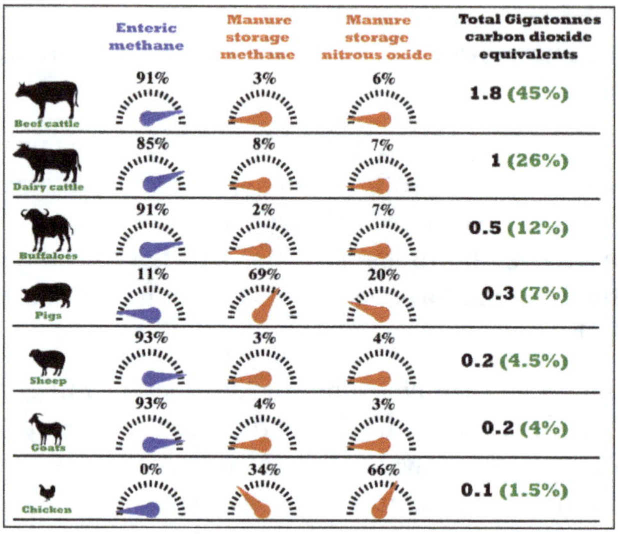

Greenhouse gases incidence of enteric fermentation and manure storage by animal type, expressed as Gigatonnes of carbon dioxide equivalents.

In addition to greenhouse gases arising from enteric fermentation and manure storage, feed production together with the related soil carbon dioxide and nitrous oxide emissions is another important hot spot for the livestock sector. Soil carbon dioxide emissions are due to soil carbon dynamics (e.g., decomposing plant residues, mineralization of soil organic matter, land use change, etc.), the manufacturing of synthetic fertilizers and pesticides, and from fossil fuel use in on-farm agricultural operations. Nitrous oxide emissions are emitted when organic and inorganic fertilizers are applied to the soil.

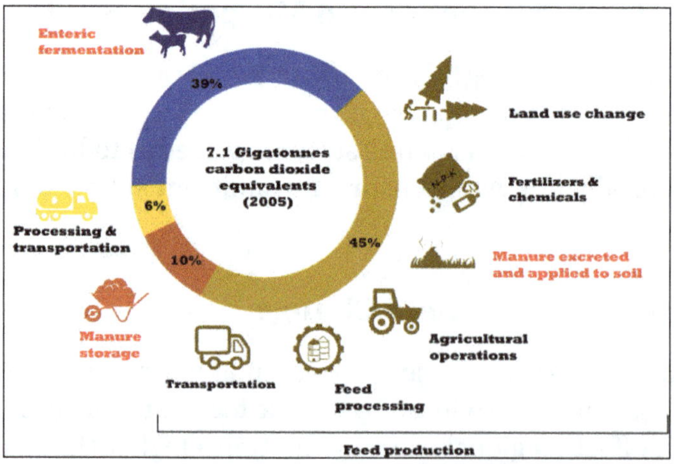

Livestock emissions by source. Direct livestock emissions are shown in red.

Feed production and processing contribute about 45% of the whole sector (3.2 Gigatonnes of carbon dioxide equivalents). Enteric fermentation producing about 2.8 Gigatonnes (39%) is the second largest source of emissions. Manure storage with 0.71 Gigatonnes accounts for about 10% of the total. The remaining 6% (0.42 Gigatonnes of carbon dioxide equivalents) is attributable to the processing and transportation of animal products. Feed production includes all the greenhouse gas emission arising from land use change, manufacturing and use of fertilizers and pesticides, manure excreted and applied to fields, agricultural operations, feed processing, and feed transport.

Enteric Fermentation

Enteric fermentation is a natural part of the digestive process of ruminants where bacteria, protozoa, and fungi contained in the fore-stomach of the animal (rumen), ferment and break down the plant biomass eaten by the animal. Plant biomass in the rumen is converted into volatile fatty acids, which pass the rumen wall and go to the liver through the circulatory system. This process supplies a major part of the energy needs of the animal and enables the high conversion efficiency of cellulose and semi-cellulose, which is typical of ruminants. The gaseous waste products of enteric fermentation, carbon dioxide and methane, are mainly removed from the rumen by eructation. Methane emission in the reticulorumen is an evolutionary adaptation that enables the rumen ecosystem to dispose hydrogen, which may otherwise accumulate and inhibit carbohydrate fermentation and fiber degradation. The emission rate of enteric methane varies according to feed intake and digestibility.

Manure Storage

Manure acts as an emission source for both methane and nitrous oxide, and the quantity emitted is linked to environmental conditions, type of management and composition of the manure. Organic matter and nitrogen content of excreta are the main characteristics influencing emission of methane and nitrous oxide, respectively. Under anaerobic conditions, the organic matter is partially decomposed by bacteria producing methane and carbon dioxide. Storage or treatment of liquid manure (slurry) in a lagoon or tank promotes an anaerobic environment which leads to an increase in methane production. Long storage periods and warm and wet conditions can further increase these emissions. On the other hand, nitrous oxide emissions need a combination of aerobic and anaerobic conditions to be produced. Therefore, when manure is handled as a solid (dung) or deposited on pastures, nitrous oxide production increases while little or no methane is emitted. Nitrous oxide is generated through both the nitrification and denitrification processes of the nitrogen contained in manure, which is mainly present in organic form (e.g., proteins) and in inorganic form as ammonium and ammonia. Nitrification occurs aerobically and converts ammonium and ammonia to nitrites and then nitrates, while denitrification occurs anaerobically converting nitrates to nitrous oxide and nitrogen gas. The balance between ammonium and ammonia is highly affected by pH, with ammonia increasing as pH increases.

Feed Production

Almost 60% of the global biomass harvested worldwide enters the livestock subsystem as feed or bedding material. Greenhouse gas emissions from feed production represent

60–80% of the emission coming from eggs, chicken and pork, and 35–45% of the milk and beef sector. As shown in Figure, emissions from feed production account for about 45% of the livestock sector. The application of manure as fertilizer for feed crops and the deposition of manure on pastures generate a substantial amount of nitrous oxide emissions representing about half of these emissions. Although livestock feed production often involves large applications of nitrogen to agricultural soils, good manure management can reduce the need for manufactured fertilizers.

Livestock Mitigation Strategies

The extreme heterogeneity of the agricultural sector needs to be taken into account when defining the overall sustainability of a mitigation strategy, which can vary across different livestock systems, species, and climates. Generally, no measure in isolation will encompass the full emission reduction potential, while a combination selected from the full range of existing options will be required to reach the best result. It is also important to consider the "pollution swapping" effect when evaluating the effectiveness of a mitigation strategy. Reduction of methane emissions during enteric fermentation might be counteracted by increased greenhouse gas emissions in applied manure. Reduction of direct nitrous oxide emissions during storage might result in higher nitrate leaching and ammonia volatilization during field application.

Mitigation may occur directly by reducing the amount of greenhouse gases emitted, or indirectly through the improvement of production efficiency. The main strategies to mitigate greenhouse gas emissions in the livestock sector have been investigated and are summarized in Table.

Table: Mitigation potential of various strategies.

Strategies	Category	Potential mitigating effect*	
		Methane	Nitrous Oxide
Enteric fermentation	Forage quality	Low to medium	†
	Feed processing	Low	Low
	Concentrate inclusion	Low to medium	†
	Dietary lipids	Medium	†
	Electrons receptors	High	†
	Ionophores	Low	†
	Methanogenic inhibitors	Low	†
Manure storage	Solid-liquid separation	High	Low
	Anaerobic digestion	High	High
	Decreased storage time	High	High
	Frequent manure removal	High	High
	Phase feeding	‡	Low
	Reduced dietary protein	‡	Medium

	Nitrification inhibitors	‡	Medium to high
	No grazing on wet soil	Low	Medium
	Increased productivity	High	High
Animal management	Genetic selection	High	‡
	Animal health	Low to medium	Low to medium
	Increase reproductive eff.	Low to medium	Low to medium
	Reduced animal mortality	Low to medium	Low to medium
	Housing systems	Medium to high	Medium to high

*High = ≥30% mitigating effect; Medium = 10–30% mitigating effect; Low = ≤10% mitigating effect. Mitigating effects refer to percent change over a "standard practice".

†Inconsistent/variable results.

‡Uncertainty due to limited research or lack of data.

Enteric Fermentation

Decreasing methane emissions from ruminants is one pressing challenge facing the ruminant production sector. Strategies for reducing this source of emissions focus on improving the efficiency of rumen fermentation and increasing animal productivity. A large number of mitigation options have been proposed (e.g., diet manipulation, vaccines, chemical additives, animal genetic selection, etc.) with different efficiencies in reducing enteric methane as shown in Table.

Forage quality and digestibility affect enteric methane production. Lignin content increases during plant growth, consequently reducing plant digestibility. Therefore, harvesting forage (especially grass) for ensiling at an earlier stage of maturity increases its soluble carbohydrate content and reduces lignification. According to Knapp et al. practices aimed to increase forage quality have shown potential enteric methane reduction of about 5% per unit of fat protein corrected milk.

Physical processing of forages, such as chopping, grinding, and steam treatment, also improves forage digestibility and mitigates enteric methane production in ruminants. However, the reduction potential of this practice was reported to be less than 2% per unit of fat protein corrected milk. Improving diet digestibility by increasing concentrate feeding is another effective mitigation strategy, reducing by 15% methane emissions per unit of fat protein corrected milk. However, the ratio of forage to concentrate has to be carefully taken into account when applying this strategy. Indeed, although a marked reduction of enteric methane can be expected with rates of concentrate inclusion between 35% and 40%. A greater proportion of dietary fermentable carbohydrates could increase the risk of metabolic diseases (e.g., rumen acidosis).

Addition of fats or fatty acids to the diets of ruminants can decrease enteric methane emissions by both decreasing the proportion of energy supplied from fermentable carbohydrates and changes in the microbial population of the rumen. Although some byproducts (e.g., cottonseed, brewer's grains, cold-pressed canola meal, etc.) are effective in reducing enteric fermentation, the mitigation potential of high oil byproducts has not been well-established and in some cases methane production may increase due to increased fiber intake. The inclusion of lipids higher than 10% can lead to impairment of ruminal function due to changes to the microbial population which in turn decreases the ability to digest fiber. Lipid diet supplementation between 5% and 8% of the dry matter intake is an effective mitigation strategy with a potential enteric methane reduction of about 15% per unit of fat protein corrected milk.

Feed additives (electron receptors, ionophoric antibiotics, chemical inhibitors, etc.) have also been tested for their ability to decrease methane emissions. However, the unknown toxicity and the health risks associated with the use of some of these compounds may severely constrain widespread adoption.

Manure Storage

Increased animal density together with continuous inflow of nutrients from imported feeds is likely to increase volumes of manure to be managed. Stored manure accounts for a relatively small amount of direct agricultural greenhouse gases, and it is technically possible to mitigate a very high percentage of these emissions. As methane production increases with the temperature of stored manure, a reduction of storage temperature has been reported to drop these emissions by 30–50%. However, the net greenhouse gas mitigation resulting from this strategy can vary widely, and it is strictly related to the energy used and the cooling system adopted.

Frequent removal of manure to an outside storage facility is an effective practice that can be accomplished using grooved floors combined with regular scraping of manure, especially for pigs and some cattle production systems. Indeed, if the channels underneath the stable are emptied regularly, and the manure/slurry are transported to an outside storage facility, this practice has the potential to reduce methane and nitrous oxide emissions by 55% and 41%, respectively. On poultry farms the litter/manure is usually removed at the end of the crop; however, advanced layer housing using belt scrapers can efficiently remove litter/manure continuously and decrease greenhouse gas emissions.

Solid-liquid separation is a processing technology that partially separates the solids from liquid manure using gravity or mechanical systems such as centrifuges or filter presses. As shown in above Table, the greenhouse gas mitigation potential of this technique has been reported to be higher than 30% compared with untreated manure. The organic component with a larger particle size follows the solid stream during the separation process, and it is then stored in stockpiles. The aerated condition of the storage can then limit the potential for methane to be emitted; however, ammonia loss through composting and generating high temperatures can be accelerated. Also, the remaining liquid fraction is still a potential source of indirect nitrous oxide emissions. Indeed, once the fibrous and large pieces of organic material are subtracted, it will not form a crust during storage, leading to increased volatilization of ammonia by increasing the mass transfer coefficient at the surface. Although greenhouse gas mitigation of the solid-liquid separation process can be partially counterbalanced by ammonia emissions, it is important to note that there are many management practices that can overcome these issues, such as covering slurry storage and the use of injection for land application.

Anaerobic digestion is a biological degradation process, which in the absence of oxygen, produces digestate and biogas (mainly methane and carbon dioxide) from manure. Biogas collected from the system is often used to generate electricity, to fuel boilers or furnaces, or to provide combined heat and power. Taking into account the greenhouse gas emissions arising from the use of the digestate as fertilizer, and the credit for the renewable energy produced, anaerobic digestion has been reported to yield more than 30% reduction in greenhouse gas emissions when compared with traditional manure handling systems. However, further attention to the management of the digestate leaving the anaerobic digestion is needed. Indeed, mineralization of the organic nitrogen occurring during biological degradation increases the inorganic nitrogen content and pH of the effluent, which in turn may increase ammonia volatilization. Combining anaerobic digestion and solid-liquid separation could reduce the amount of ammonia lost following digestion.

Diet severely affects excretion of nitrogen in most farm animals; therefore grouping livestock on the basis of their feed requirements can help in reducing this source of nitrous oxide in the excreta. Although a low-protein diet could effectively mitigate nitrous oxide emissions from cattle manure storage, some attention must be given to manipulating dietary nitrogen. For example, decreasing protein could lead to an increase of fermentable carbohydrates, which in turn will likely increase methane production.

The diet for all animal species should be balanced for amino acids to avoid a depression in feed intake and a decrease in animal productivity. Manufactured amino acids are routinely used to balance the diet of monogastrics (pigs and poultry), but the environmental impact associated with the manufacturing of these supplements must be considered when including amino acids as a greenhouse gas mitigation strategy. In ruminants, supplementation of free amino acids results in fast degradation in the rumen, without a significant increase in animal productivity. On the contrary, rumen-protected amino acids resist chemical alterations in the rumen and can reach the intestine where they are absorbed, improving milk yield in dairy cows. Overall, feeding protein close to the animal's requirement is recommended as an effective mitigation strategy to reduce ammonia and nitrous oxide emissions from manure.

Feed Production

The timing, quantity, and method of fertilizer applications are important factors influencing soil nitrous oxide emissions. The nitrogen fertilizer applied is susceptible to loss by leaching and denitrification before crop uptake. Therefore, ensuring that appropriate amounts of nitrogen get to the growing crop and avoiding application in wet seasons or before major rainfall events, are valuable practices which could help in optimizing biomass production and reduce soil greenhouse gas emissions.

As lower methane emissions occur after manure land application, decreasing storage time can effectively help in reducing greenhouse gas emissions. However, the resulting frequent soil applications can have a variable effect on nitrous oxide emissions from field and carbon dioxide emissions from fuel combustion. Avoiding application during prolonged periods with wet soil and periods of low plant nitrogen uptake could help in increasing the effectiveness of this practice.

Adequate storage facilities can provide greater flexibility in choosing when to apply manure to fields, while the use of on-farm manure analysis could help the farmer develop a nutrient management plan and minimize environmental impacts. The use of nitrification inhibitors has the potential to reduce nitrogen leaching by inhibiting the conversion of ammonia to nitrate. However, this beneficial effect is weakened by a reported increase in indirect nitrous oxide emission that can result from increased ammonia volatilization. This highlights the importance of considering both gases when evaluating the use of nitrification inhibitors as an option to mitigate climate change. Overall, nitrification inhibitors have been demonstrated as an effective practice to reduce nitrous oxide emissions.

Intensive rotational grazing systems are being promoted as a good way to increase forage production and reduce nitrous oxide emissions. These systems are characterized by multiple smaller fields called paddocks for the rotation of livestock. By subdividing pastures and rotating animals, farmers can manage stocking densities and grazing duration and thereby manage nitrogen excreta distribution and vegetation regrowth. A more uniform distribution of urine throughout the paddock would reduce the effective nitrogen application rate, which could translate into a reduction in nitrous oxide emissions. Keeping animals off the paddocks during wet weather will reduce sward damage and soil compaction. In addition, avoiding excreta deposition at these times will reduce nitrous oxide emissions and nitrogen leaching.

Animal Management

There is a direct link between greenhouse gas emission intensities and animal efficiency. The more productive the animal is, the lower the environmental impact will be (on as per unit of product basis). Both management quality and expression of full genetic potential are necessary to increase production efficiency.

Breeding for more productive animals can lead to a reduction of the nutrient requirements needed to reach the same level of production. This is a valuable greenhouse gas mitigation strategy. A more efficient animal will retain more dietary nitrogen protein and there will less nitrogen in feces and urine. Genetic improvement of daily gain and feed conversion that has been achieved in broilers over the last 20 years has reduced substantially the emissions per unit of weight. Nevertheless, strategies that aim to change animal phenotypes to enhance productivity or efficiency may harm animal health and welfare unless these effects are measured and controlled. Animals of a particular genotype selected for increased production will only be able to realize this potential on a high input system in which resources are adequately supplied. In other words, new breeds and crosses can lead to substantial greenhouse gas reduction, but they need to fit within production systems and climates that may be characterized by limited resources and other constraints.

Poor fertility means that more breeding animals are required in the herd to meet production targets, and more replacements are required to maintain the herd size, which in turn increases greenhouse gas emissions. Improved fertility in dairy cattle could lead to a reduction in methane emissions by 10–24% and reduced nitrous oxide by 9–17%. Nevertheless, increasing reproductive pressure may increase the metabolic demands associated with pregnancy and lactation that could negatively affect animal health and increase the risk of metabolic diseases, reduce immune function and in turn reduce fertility. Poorer livestock health and welfare are associated with behavioral and metabolic changes, which can effect greenhouse gas emissions in several ways. Animals fighting an infection will need more energy for maintenance.

Animal Welfare

Animal welfare refers to the quality of life experienced by an animal and encompasses how well the animal is coping with his or her current situation and surroundings. World Animal Protection believes animal welfare is affected by the relationships human beings have with animals and it is our duty to ensure all animals are treated humanely, responsibly, and with respect. This includes consideration for all aspects of animal well-being, such as the Five Domains. Animal welfare should be at the forefront of every human action involving animals.

Five Domains of Animal Welfare

- Nutrition: Factors that involve the animal's access to sufficient, balanced, varied, and clean food and water.
- Environment: Factors that enable comfort through temperature, substrate, space, air, odour, noise, and predictability.
- Health: Factors that enable good health through the absence of disease, injury, impairment with a good fitness level.

- Behaviour: Factors that provide varied, novel, and engaging environmental challenges through sensory inputs, exploration, foraging, bonding, playing, retreating, and others.
- Mental State: By presenting positive situations in the previous four functional domains, the mental state of the animal should benefit from predominantly positive states, such as pleasure, comfort, or vitality while reducing negative states such as fear, frustration, hunger, pain, or boredom.

Why is Animal Welfare Important?

Animal welfare is important because there are so many animals around the world suffering from being used for entertainment, food, medicine, fashion, scientific advancement, and as exotic pets. Every animal deserves to have a good life where they enjoy the benefits of the Five Domains.

References

- Animal-husbandry, biology: vedantu.com, Retrieved 12, January 2020
- Current-Scientific-Issues: oie.int, Retrieved 07, June 2020
- Organic-animal-husbandry-Breeding-housing-and-feeding-animals, AnimalHealth: infonet-biovision.org, Retrieved 13, February 2020
- The-innovative-techniques-in-animal-husbandry, animal-husbandry-and-nutrition: intechopen.com, Retrieved 10, July 2020
- Important-livestock-management-tips-for-farmers, animal-husbandry: krishijagran.com, Retrieved 09, March 2020
- What-is-livestock-management-software: navfarm.com: Retrieved 26, August 2020
- Animal-welfare-and-why-important: worldanimalprotection.org.nz, Retrieved 18, April 2020
- Nutrition-and-health-management-in-dairy-production, livestock-health-and-farming: intechopen.com, Retrieved 25, May 2020

2

Animal Nutrition, Feeding and Health

Animal nutrition involves feed preparation and focuses on the dietary nutrition needs of animals. Animal feed is the food given to domestic animals. It includes straw, hay, silage, oils and mixed rations, compressed and pelleted feeds, etc. The chapter closely examines the key concepts of animal nutrition, feeding and health to provide an extensive understanding of the subject.

Animal Nutrition

Food consumed by animals gets digested, absorbed and used to generate energy to perform day to day activities. But not all the components of food are digestible. The components of food that are utilized by the body are called nutrients and the process of obtaining the food needed for health and growth is known as nutrition.

Classes and Sources of Nutrition

The six major classes of nutrients are water, protein, carbohydrates, fats, minerals and vitamins. There are various sources of nutrition. In the case of animals, roughages and concentrates are the main sources of nutrition. Roughages, exemplified by grasses, roughages and silages are rich in fibre and low in energy. On the other hand, concentrates such as grains are rich in energy and low in fibre.

- Water constitutes more than one-half of the animals' bodies. It is vital for all physiological processes such as digestion, blood circulation and waste elimination. Water is the medium in which all the nutrients are dissolved. Water regulates body temperature through perspiration and evaporation. It is important to supply fresh and clean water to animals.
- Protein is essential for healthy growth and maintenance and reproduction. Amino acids, the building blocks of protein are essential for the formation of tissues. Protein can be of plant or animal origin. Livestock feeds majorly consist of plant proteins such as oil cake meals, hay and straws. Fish meal, meat scraps etc. are sources of animal protein for livestock.
- Carbohydrates are the main source of energy for the regular functioning of the body. The main source of carbohydrates in livestock feed are grains such as oats, wheat, barley, corn, sorghum etc. forages and hay.

- Fats are an important part of the animal diet; nevertheless, they are needed in small amounts. Fats act as storehouses of energy and also insulate the body. Oilseeds are the main source of fats in agriculture feeds.
- Minerals aid in the healthy maintenance of bones and teeth, tissues and also regulate body chemical processes. Minerals are categorized as microminerals which are required in smaller amounts and macro minerals that are required in larger amounts by the body. Mineral intake accounts for a smaller proportion of the diet and their excess intake would result in toxicity.
- Like minerals, vitamins are also required in smaller amounts. They perform a wide array of functions like metabolism, growth and reproduction, red blood cell maturation, digestion, bone and teeth formation, combating stress and prevention of infection. Vitamins can either be water-soluble or fat-soluble. Broadly, they are classified as vitamin A, B, C, D, E and K.

Desirable Characteristics of Nutrition

- The quantity of the feed should meet the requirements and production of the animal.
- The nutrition provided should be palatable.
- Balanced nutrition is the key to healthy growth. There should be a better and balanced mixture of proteins, vitamins and other nutrients.
- The mineral matter should be present in enough quantities.
- The feed should be fairly laxative to avoid constipation and other digestive troubles in the animals.
- The feed should be fairly bulk.
- Regularity in the feed timings should be maintained.
- There should be enough green succulent fodder as they are bulky in nature, nutritious and easily digestible.
- Sudden changes in the feed should be avoided.
- It should be economical in terms of labour and cost.

Dietary Sources and their Effects on Animal Production and Environmental Sustainability

Animal agriculture has been an important component in the integrated farming systems in developing countries. It serves in a paramount diversified role in producing animal protein food, draft power, farm manure as well as ensuring social status-quo and enriching livelihood. Ruminants are importantly contributable to the well-being and the livelihood of the global population. Ruminant production systems can vary from subsistence to intensive type of farming depending on locality, resource availability, infrastructure accessibility, food demand and market potentials.

The growing demand for sustainable animal production is compelling to researchers exploring the potential approaches to reduce greenhouse gases (GHG) emissions from livestock. Global warming has been an issue of concern and importance for all especially those engaged in animal agriculture. Methane (CH_4) is one of the major GHG accounted for at least 14% of the total GHG with a global warming potential 25-fold of carbon dioxide and a 12-year atmospheric lifetime. Agricultural sector has a contribution of 50 to 60% methane emission and ruminants are the major source of methane contribution (15 to 33%). Methane emission by enteric fermentation of ruminants represents a loss of energy intake (5 to 15% of total) and is produced by methanogens (archae) as a result of fermentation end-products.

Ruminants' digestive fermentation results in fermentation end-products of volatile fatty acids (VFA), microbial protein and methane production in the rumen. Rumen microorganisms including bacteria, protozoa and fungal zoospores are closely associated with the rumen fermentation efficiency. Besides using feed formulation and feeding management, local feed resources have been used as alternative feed additives for manipulation of rumen ecology with promising results for replacement in ruminant feeding. Those potential feed additive practices are as follows:

- The use of plant extracts or plants containing secondary compounds (e.g., condensed tannins and saponins) such as mangosteen peel powder, rain tree pod.
- Plants rich in minerals, e.g., banana flower powder.
- Plant essential oils, e.g., garlic, eucalyptus leaf powder, etc. Implementation of the -feed-system using cash crop and leguminous shrubs or fodder trees is of promising results.

Livestock production is undertaken in a multitude of ways across the planet, providing a large variety of goods and services, and using different animal species and different sets of resources, in a wide spectrum of agro-ecological and socio-economic conditions. Global livestock systems occupy about 30% of the planet's ice-free terrestrial surface area and are a significant global asset with a value of at least $1.4 trillion. Currently, livestock is one of the fastest growing agricultural subsectors in developing countries. This growth is driven by the rapidly increasing demand for livestock products, this demand being driven by population growth, urbanization and increasing incomes in developing countries. This combination of growing demand in the developing world and stagnant demand in industrialized countries represents a major opportunity for livestock keepers in developing countries, where most demand is met by local production, and this is likely to continue well into the foreseeable future. Along with an exploration of food consumption trends and projections to 2050, both globally and for different regions of the world, the drivers largely responsible for these observed consumption trends will be examined. At the same time, the expansion of agricultural production needs to take place in a way that allows the less well-off to benefit from increased demand and that moderates its impact on the environment.

Although integral to many farming systems, livestock production is nevertheless associated with many impacts that are deemed socially undesirable. Whereas animal welfare concerns have been documented for centuries, damage attributed to and responsibility for greenhouse gas (GHG) emissions are more recent concerns. Enteric methane (CH_4) emission in ruminants, which is produced via fermentation of feeds in the rumen and lower digestive tract by methanogenic archaea, represents a loss of 2 to 12% of gross energy of feeds and contributes to global greenhouse effects. Globally, about 80 million tonnes of CH_4 is produced annually from enteric fermentation mainly from ruminants. Therefore, CH_4 mitigation strategies in ruminants have been focused on obtaining economic as well as environmental benefits.

Global Animal Production Systems

Production environments, and the intensities and purposes of production, vary greatly within and across countries. Animal production systems have been categorised on the basis of agro-ecological opportunities and demand for livestock commodities. In many of these systems, the livestock element is interwoven with crop production, as in the rice/buffalo or cereal/cattle systems of Asia. Animal manure is often essential for maintaining soil fertility, and the role of animals in nutrient cycling is often an important motivation for keeping animals, particularly where this involves a transfer of nutrients from common property resources to private land. Many of these systems that are the result of a long evolution are currently under pressure to adjust to rapidly evolving socio-economic conditions; large intensive livestock production units, in particular for pig and poultry production, have emerged over the last decades in many developing regions in response to the rapidly growing demand for livestock products.

Moreover, the degree to which each system is integrated into the market economy varies according to a host of factors, perhaps the most important of which is geographical location. The influence of geographical location on market integration is twofold: partly agro-ecological and partly infrastructural. Some areas may have a higher degree of market integration because rainfall and soil conditions are conducive to cash cropping and the production of surpluses; others may lack one or both of these advantages but are compensated by their relative proximity to urban markets and other facilities. Animal production systems can be described into categories as follows:

Subsistence Animal Production Systems

For the subsistence-oriented household, land and labour are the principal factors of production. Capital investment is limited to non-monetary self-produced equipment, land improvement and livestock raised through natural reproduction. Increases in production are mainly dependent on the weather and on the quantity and quality of those factors of production controlled by the household. These, for example, may include:

- Use of surplus labour for bush clearing and erosion control.
- Use of animal manure to raise soil fertility.
- Better livestock management practices.

Progress in production is likely to be slow but improvements are possible through farming systems research, education and extension programmes. There are few local off-farm employment opportunities. The monetary circuit plays little role in the economy of the mainly subsistence-oriented household. For the subsistence-oriented farm, output and consumption are identical. Such households thus remain largely (but not wholly) unresponsive to price and market signals. Families living under these conditions rarely aim to maximize production, since this would imply specialization, with its attendant risks. Rather, the goal is to maximize the chances of survival. A mainly subsistence-oriented farmer will be reluctant to shift from a traditional practice to a new technology if doing so incurs greater risk of failure.

Semi-subsistence Animal Production Systems

A semi-subsistence household produces a considerable proportion of its consumption requirements (60 to 80%). In addition, it will produce cash crops such as vegetables, coffee and tea, and keep livestock for sale. The semi-subsistence producer will therefore be confronted with the risks associated with price fluctuations and with variations in the natural environment. The monetary circuit thus assumes an important role in the semi-subsistence production unit. Such units tend to be more responsive to market and price signals than the subsistence-oriented producers. The higher the share of output being sold on the market, the greater the importance of the monetary circuit in the semi-subsistence production system. The impact of market and price signals will ultimately depend on the degree of market integration.

What are the reasons behind a household's desire to enter the monetary circuit? Answering this question will help us understand the factors which influence production responses. The first step in the transition process from subsistence to more commercialized production may be a need to obtain cash to meet legal or social obligations, such as the payment of school fees or the hosting of a wedding reception. Insofar as such needs are the only purpose of sales, there will be a negative relationship between price and market supply. In other words, the higher the market price, the smaller will be the amounts that need to be sold and vice versa.

As the transition process continues, market supply responses become positive as producers recognize that increasing their cash income enables them to buy other consumer goods which improve their welfare. If these goods are regularly available at local markets, income growth may become an important family goal. Higher income also enables a household to purchase more external inputs (fertilizer, seeds, etc.), thus increasing output still further in the future. Finally, cash can also be used to pay interest and principal on credit, opening up greater opportunities for investment and hence the development of new enterprises. Thus, the transition from pure subsistence, through semi-subsistence to more commercial farming will have two interrelated effects on consumption and production in the rural household, namely:

- The direct acquisition of consumer goods and services.
- The further growth of income through increased use of external inputs.

For families living under these systems, risk aversion remains an important determinant of household decisions. These producers confront the risks associated with price fluctuation as well as those resulting from climate. Sometimes these will offset one another, as when low yields lead to scarcity, causing market prices to rise, and vice versa. At other times, factors bearing no relationship to yield variations will influence prices. For semi-subsistence producers, innovations with minimal input of external factors of production could be offered.

Intensive (Commercial) Animal Production Systems

In these systems, the monetary circuit becomes more important than the physical one, which may become less complex as a result of specialization. These production units tend to be highly responsive to price and market signals, switching enterprises and increasing or decreasing their market involvement in accordance with them. Increases in production are almost certain to involve the use of external inputs and services. Progress in production can be rapid, but dramatic setbacks may occasionally occur. Off-farm employment opportunities are more common and are found nearer home. For families living under these conditions, the allocation of resources will be determined largely by the profit rather than the survival motive. However, although risk aversion plays a smaller part in decision making, households will tend to refrain from fully commercial production if markets are unreliable or if institutional support (access to credit, price stabilization schemes, animal health services, etc.) is inadequate.

The presence of commercial systems is connected to both demand factors and supply determinants; areas with high population density and purchasing power, in particular coastal areas in East Asia, Europe and North America, which also have access to ocean ports, show a high prevalence of industrial systems and import much of the necessary feed. In contrast, there are areas with ample feed supplies such as the mid-western United States of America (USA) and interior parts of Brazil and Argentina, where industrial systems rely mainly on local feed surpluses. East and Southeast Asia strongly dominate industrial monogastrics' production in the developing regions. Southern Brazil is another industrial production hot spot at world level, while important regional centres of industrial production are found, for example in Mexico, Colombia, Venezuela and Chile. Similarly there are major regional centres for the industrial production of chicken in Nigeria, South Africa and the Middle East.

Animal Protein Products Consumption Demand for Increasing Global Population

There has been an increasing pressure on the livestock sector to meet the growing demand for high-value animal protein. The world's livestock sector is growing at an unprecedented rate and the driving force behind this enormous surge is a combination of population growth, rising incomes and urbanization. Annual meat production is projected to increase from 218 million tonnes to 376 million tonnes by 2030. There is a strong positive relationship between the level of income and the consumption of animal protein, with the consumption of meat, milk and eggs increasing at the expense of staple foods. Because of the recent steep decline in prices, developing countries are embarking on higher meat consumption at much lower levels of gross domestic product than the industrialized countries did about 20 or 30 years ago.

Urbanization is a major driving force influencing global demand for livestock products. Urbanization stimulates improvements in infrastructure, including cold chains, which permit trade in perishable goods. Compared with the less diversified diets of the rural communities, city dwellers have a varied diet rich in animal proteins and fats, and characterized by higher consumption of meat, poultry, milk and other dairy products. Table shows trends in per capita consumption of livestock products in different regions and country groups. There has been a remarkable increase in the consumption of animal products in countries such as Brazil and China, although the levels are still well below the levels of consumption in North American and most other industrialized countries. Consumption of meat in the U.S. was highest when compared to the global average. The countries that consume the least amount of meat are in Africa and South Asia; the lowest ten are Sierra Leone, Democratic Republic of Congo, Mozambique, Sri Lanka, Rwanda, India, Malawi, Guinea, Burundi and Bangladesh. Consumption in these countries is between 3 and 5 kg per capita per year. This is compensated to some extent in Bangladesh by higher fish consumption (17.5 kg) and in India and Sri Lanka by higher milk consumption (47.5 and 35.9 kg, respectively). Milk consumption in the U.S. is 118 kg per capita per year. Many African countries are in the bottom quartile for consumption of meat plus fish combined.

Table: Protein consumption demand per capita of livestock products.

Region	Meat, kg/yr			Milk, kg/yr		
	1964 to 1966	1997 to 1999	2030	1964 to 1966	1997 to 1999	2030
World	24.2	36.4	45.3	73.9	78.1	89.5
Developing countries	10.2	25.5	36.7	28.0	44.6	65.8
Near East and North Africa	11.9	21.2	35.0	68.6	72.3	89.9

Sub-Saharan Africa	9.9	9.4	13.4	28.5	29.1	33.8
Latin America and the Caribbean	31.7	53.8	76.6	80.1	110.2	139.8
East Asia	8.7	37.7	58.5	3.6	10.0	17.8
South Asia	3.9	5.3	11.7	37.0	67.5	106.9
Industrialized countries	61.5	88.2	100.1	185.5	212.2	221.0
Transition countries	42.5	46.2	60.7	156.6	159.1	178.7

As diets become richer and more diverse, the high-value protein that the livestock sector offers improves the nutrition of the vast majority of the world. Livestock products not only provide high-value protein but are also important sources of a wide range of essential micronutrients, in particular minerals such as iron and zinc, and vitamins such as vitamin A. For the large majority of people in the world, particularly in developing countries, livestock products remain a desired food for nutritional value and taste. Excessive consumption of animal products in some countries and social classes can, however, lead to excessive intakes of fat.

Greenhouse Gases and Animal Contribution

The growing demand for livestock products is likely to have an undesirable impact on the environment. For example, there will be more large-scale, industrial production, often located close to urban centres, which brings with it a range of environmental and public health risks. Environmental impacts of livestock production have historically been confined to more localized problems of overgrazing, desertification, and water pollution by poor waste handling. Such concerns were often offset by recognition of the cultural significance of livestock and more tangible benefits from the use of animal products and manures in farming systems. In developing countries, livestock production provides not only food, but also a wide range of non-food benefits including income, employment, and many other contributions to rural and social development. The need to respond to global climate change has focused attention on the main sources of emissions with all significant sources coming under scrutiny. This is largely because developed countries have committed themselves to externally defined emissions reductions (mitigation) targets that must somehow be shared amongst polluting industries within their jurisdictional control.

Livestock production systems contribute an estimated 18% of global anthropogenic GHG emissions. These emissions represent a significant proportion for some countries, including New Zealand, Ireland, and the United Kingdom. The main sources and types of GHG from livestock systems are methane production from animals (25%), carbon dioxide (CO_2) from land use and its changes (32%), and nitrous oxide (N_2O) from manure and slurry management (31%).

Strategies in Preventing and Alleviating Greenhouse Gases Especially from Animals

Livestock are already well-known to contribute to GHG emissions and accounting for about 18% of the anthropogenic GHG emissions. Among domesticated livestock, ruminant animals (cattle, buffalo, sheep, goats, and camels) produce significant amounts of CH_4 and they are produced in the rumen and hind gut. Fermentation of feeds in the rumen is the largest source of CH_4 from enteric fermentation. Methane production from ruminant is a complex process that involves a group of Archaea known collectively as methanogens, which belong to the phylum Euryarcheota. During this process, rumen microbes convert ingested organic matter into energy for microbial growth, and into fermentation end-products, including VFA, alcohols, H_2, and CO_2. The major VFA produced in the rumen include acetate (Ac), propionate (Pr) and butyrate (Bu), which generally account for more than 95% of the total VFA production. Excess reducing power generated during conversion of hexose to Ac or Bu is utilized in part by Pr, but mainly by conversion to CH_4.

Methanogenic archaea are able to take some of these end products and reduce them with H_2 to produce CH_4 and H_2O. Accumulated CH_4 and other volatile gases produced in the rumen are eventually expelled through the mouth into the atmosphere via eructation. Methane emissions in ruminants accounts for 2 to 12% of gross energy loss of feeds depending upon the type of diets. Moreover, Leng reported that CH_4 emitted by the world's farmed ruminant livestock accounts for about one quarter of all anthropogenic CH_4 emission, typically estimated at 80 to 90 Tg/yr (1 Tg = 1 million tonnes) of a total of around 350 Tg/yr. Therefore, lowering global CH_4 emissions from enteric fermentation is an important part of any effort to reduce anthropogenic GHG emissions and ruminant producers are also seeking to identify and promote good management practices. The current approaches for reducing methane production from ruminants are shown in Table.

Table: Methane abatement strategies, mechanism of abatement, considerations for use and reducing efficiency of CH_4[1].

CH_4 abatement strategies	Mechanism of abatement	Considerations for use	Reducing efficiency of CH_4
Feed management: Roughage to concentrate ratio, increased hemicellulose/starch, reducing cell wall.	Increased rate of passed; increased C_3 to C_2 ratio, reduced rumen pH.	Shift methanogensis is to hind gut or manure, risk of subacute ruminal acidosis.	7 to 90%
Plant Compounds: Condensed tannins, saponins, essential oils, organosulfur compound.	Antimicrobial activity; reduced H availability.	Optimum dosage unknown; more in vivo research needed; long-term studies needed; may affect digestibility; residues unknown.	10 to 96%

Organic Acids: Fumarate, malate nitroethane, nitrate, thiamine, bromochloro-methane.	H sink, greater proportion propionate versus acetate: Thiamine: Inhibits pyruvate oxidative decarboxylation. Bromochloromethane inhibits cobamide de-pendent methyl in CH_4 production.	Varies with diet; more in vivo research needed; long-term studies needed; may affect digestibility.	3 to 75%
Ionophore: Monensin or rumensin.	Inhibits protozoa and gram-positive bacteria; lack of substrate for methanogenesis.	Adaptation of mi-crobiota may occur; varies with diet and animal; banned in the European Union.	4 to 76%
Immunisation and bio-logical control: Methanogen vaccine, methanotrophs, pro-biotic, bacteriophages, bacteriocins.	Host immune response to methanogens.	Vaccine targets; diet and host geograph-ical location differ-ences .	7 to 50%
Defaunation: Chemical, feed additives.	Removes associated methanogens; less H for methanogenesis.	Adaptation of mi-crobiota may occur; varies with diet; maintenance of de-faunated animals.	20 to 60%
Lipids: Fatty acids, oils, seed oils, tallow.	Inhibition of metha-nogens and protozoa; greater proportion propi-onate versus acetate; bio hydrogenation.	Effect on palatability, intake, performance, and milk compo-nents; varies with diet and ruminant species; long-term studies needed.	10 to 90%
Genetic selection	Genetic selection of animals for decreasing methane emissions.	Varies with diet and ruminant species; long-term studies needed.	

Feed Management

It is well established that increasing the level of concentrate in the diet leads to a reduction in CH_4 emissions as a proportion of energy intake or expressed by unit of animal product (milk and meat). A meta-analysis of the bibliography showed that the relationship between concentrate proportion in the diet and CH_4 production is curvilinear. Methane losses appear relatively constant (6 to 7% of GE intake) for diets containing 30 to 40% concentrate and then decrease rapidly to low values (2 to 3% of GE intake) for diets containing 80 to 90% concentrate. Replacing structural carbohydrates from forages (cellulose, hemicellulose) in the diet with non-structural carbohydrates (starch and sugars) contained in most energy-rich concentrates is associated with increases in-feed intake, higher rates of ruminal fermentation and accelerated feed turnover, which results in large modifications of rumen physico-chemical conditions and microbial populations.

A shift of VFA production from acetate towards propionate occurs with the development of starch-fermenting microbes. This results in a lower CH_4 production because the relative proportion of ruminal hydrogen sources declines whereas that of hydrogen sinks increases. However, this low acetate to propionate ratio may not be always observed in high-concentrate fed animals, that is, young bulls fed maize grain-based diets containing 30 or 45% starch had a similar ratio (2.50 vs. 2.88, respectively). The lower CH4 emissions from bulls fed the diet containing 45% starch compared to those fed other two diets containing 30% starch (2.5% vs. 6.9% of GE intake, respectively) could be better explained by a lower ruminal pH (5.06 vs. 5.90, respectively) and a decrease in protozoal number. The low ruminal pH might also inhibit the growth and/or activity of methanogens and of cellulolytic bacteria. A positive correlation between cellulolytic bacteria and methanogens in the rumen of different species (cattle, sheep, llamas, deer) has been shown, except in buffalos. This exception was explained by the fact that Fibrobacter succinogenes, a non-hydrogen-producing cellulolytic species, was the major cellulolytic bacteria of this animal species.

Concerning the effect of the nature of concentrate on methanogenesis, few direct comparisons have been carried out. Concentrates rich in starch (wheat, barley, maize) have a more important negative effect on CH_4 production than fibrous concentrates (beet pulp). Substitution of beet pulp by barley in a high concentrate diet (70%) fed to dairy cows reduced CH_4 emissions by 34%. Lovett et al. reported that this was not the case when fresh forages were the main ingredients of the basal diet. Beauchemin et al. measured CH_4 emissions from feedlot cattle fed backgrounding and finishing diets containing maize (slowly degradable starch) or barley grain (rapidly degradable starch). Effect of grain source on CH_4 emissions was conditioned by the production phase. Expressed on the basis of GE intake, CH_4 emissions during the backgrounding phase were not affected by grain source, whereas emissions were surprisingly less for the maize finishing diet than for the barley finishing period.

Animal Feed

Feed, also called animal feed is the food grown or developed for livestock and poultry. Modern feeds are produced by carefully selecting and blending ingredients to provide highly nutritional diets that both maintain the health of the animals and increase the quality of such end products as meat, milk, or eggs. Ongoing improvements in animal diets have resulted from research, experimentation, and chemical analysis by agricultural scientists.

Silo and auger feeding trough.

Animals in general require the same nutrients as humans. Some feeds, such as pasture grasses, hay and silage crops, and certain cereal grains, are grown specifically for animals. Other feeds, such as sugar beet pulp, brewers' grains, and pineapple bran, are by-products that remain after a food crop has been processed for human use. Surplus food crops, such as wheat, other cereals, fruits, vegetables, and roots, may also be fed to animals. History does not record when dried roughage or other stored feeds were first given to animals. Most early records refer to nomadic peoples who, with their herds and flocks, followed the natural feed supplies. When animals were domesticated and used for work in crop production, some of the residues were doubtless fed to them.

The first scientific effort to evaluate feeds for animals on a comparative basis was probably made in 1809 by the German agriculturist Albrecht von Thaer, who developed "hay values" as measures of the nutritive value of feeds. Tables of the value of feeds and of the requirements of animals in Germany followed and were later used in other countries.

Preservation of green forages such as beet leaves and corn (maize) plants by packing them in pits in the earth has long been practiced in northern Europe. The idea of making silage as a means of preserving and utilizing more of the corn plant was gradually developed in Europe and was taken from France to the United States in the 1870s. When the mature, dried corn plant was fed to cattle in the winter, much of the coarse stem was wasted, but when it was chopped and ensiled (made into silage), everything was eaten. During the 20th century, concrete bunker silos for storage of silage became a common sight in many rural areas worldwide.

Basic Nutrients and Additives

The basic nutrients that animals require for maintenance, growth, reproduction, and good health include carbohydrates, protein, fat, minerals, vitamins, and water. The energy needed for growth and activity is derived primarily from carbohydrates and fats. Protein will also supply energy, particularly if carbohydrate and fat intake is inadequate or if protein intake exceeds the needs of the body. Animals need a source of energy to sustain life processes within the body and for muscular activity. When the energy intake of an animal exceeds its requirements, the surplus is stored as body fat, which can be utilized later as a source of energy if less food becomes available.

Proteins

For immature animals, protein is also needed for growth of the muscles and other parts of the body. Since milk, eggs, and wool contain much protein, additional amounts are needed in the feed of animals producing these. All animals require a small amount of protein for maintenance–i.e., the daily repair of muscles, internal organs, and other body tissues.

Proteins are composed of more than 20 different amino acids, which are liberated during digestion. Animals with a simple single stomach (monogastric), including humans, monkeys, swine, poultry, rabbits, and mink, require correct amounts of the following 10 essential amino acids daily: arginine, histidine, isoleucine, leucine, lysine, methionine, phenylalanine, threonine, tryptophan, and valine. In addition to these, poultry need glycine and glutamic acid for growth. Cystine can replace up to half of the methionine requirement, and tyrosine can replace up to half of the phenyalanine requirement. High-quality protein, such as that supplied by eggs, milk, fish meal, meat by-products, and soybean meal, contains high concentrations of the essential amino acids in the proper balance for their full utilization. Poor-quality protein, such as that in most grains, including corn, barley, and sorghum, contains too little of one or more essential amino acids. Feeds having poor-quality proteins are useful when blended with other feeds that restore the balance in essential amino acids.

A protein source's amino acid profile is of secondary importance to ruminants, such as cattle, sheep, goats, and the other animals that have four stomachs, because the bacteria that aid in the digestion of food in the rumen (first stomach) use simple nitrogen compounds to build proteins in their cells. Further on in the digestive tract, the animals digest the bacteria. By this indirect means, ruminants produce high-quality protein from a food that might originally have contained poor protein or from urea (a nitrogen compound). Very young ruminants, such as calves, lambs, and kids, however, need good-quality protein until the rumen develops sufficiently for this bacterial process to become established.

Carbohydrates and Fats

Most animals get energy from carbohydrates and fats, which are oxidized in the body. These yield heat, which maintains body temperature, furnishes energy for growth and muscle activity, and sustains vital functions. Animals need much more energy (and more total feed) for growth, work, or milk production than for simple maintenance. Simple carbohydrates such as sugars and starches are readily digested by all animals. The complex carbohydrates (cellulose, hemicelluloses) that make up the fibrous stems of plants are broken down by bacterial and protozoal action in the rumen of cattle and sheep or in the cecum of rabbits and horses. Such complex carbohydrates cannot be digested by humans or, to any appreciable extent, by dogs, cats, birds, or laboratory animals. Thus, ruminants and some herbivorous animals obtain much more of the energy-giving nutrients from the carbohydrates of plants than do monogastric carnivores and omnivores, for which fibrous materials have little or no energy value.

Fat in feeds has a high nutritive value because it is easily digested and because it supplies about two and one-quarter times as much energy as an equal weight of starch or sugar. While fat has a high nutritive value, it can be replaced by an equivalent amount of digestible carbohydrates in the feed, except for small amounts of essential fatty acids. Very small amounts of the unsaturated fatty acid linoleic, contained in some fats, are necessary for growth and health. Animal feeds typically supply ample amounts of this acid unless it has been removed by processing.

Minerals

Minerals essential for animal life include common salt (sodium chloride), calcium, phosphorus, sulfur, potassium, magnesium, manganese, iron, copper, cobalt, iodine, zinc, molybdenum, and selenium. The last six of these can be toxic to animals if excessive amounts are eaten.

All farm animals generally need more common salt than is contained in their feeds, and they are supplied with it regularly. Of the other essential minerals, phosphorus and calcium are most apt to be lacking, because they are heavily drawn upon to produce bones, milk, and eggshells. Good sources of calcium and phosphorus are bonemeal, dicalcium phosphate, and defluorinated phosphates. Eggshells are nearly pure calcium carbonate. Calcium may readily be supplied by ground limestone, ground seashells, or marl, which is all high in calcium.

Small amounts of iodine are needed by animals for the formation of thyroxine, a compound containing iodine, secreted by the thyroid gland. A serious deficiency of iodine may cause goitre, a disease in which the thyroid gland enlarges greatly. In certain regions, goitre has caused heavy losses of newborn pigs, lambs, kids, calves, and foals. Iodine deficiencies can be prevented by supplying iodized salt to the mother before the young are born. Almost all commercial sources of salt for animals contain iodine as a routine additive.

In some areas, soil and forage are deficient in copper and cobalt, which are needed along with iron for the formation of hemoglobin, the oxygen-carrying pigment of the red blood cells. In these areas, farm animals may suffer from anemia unless the deficiency is corrected by means of a suitable mineral supplement. Ruminants are usually fed cobalt in the diet so they can then synthesize vitamin B_{12}. Monogastric animals, such as pigs, require a direct source of vitamin B_{12} in their diet.

Iron, used in hemoglobin formation, is amply supplied in most animal feeds, except milk. The only practical problem with iron deficiency occurs in young suckling pigs before they start to consume other feeds in addition to milk. They require an iron injection or access to fresh soil to meet their iron needs.

Manganese is essential for animals, and the usual diets for all farm animals supply sufficient quantities. A lack of manganese may cause the nutritional disease of chicks and young turkeys called slipped tendon (perosis) and also may cause failure of eggs to hatch. Normal diets for swine are often deficient in zinc, especially in the presence of excess calcium. Fortifying feed by adding 100 parts per million of zinc, as zinc sulfate or zinc carbonate, prevents zinc deficiency symptoms, which include retarded growth rate and severe scaliness and cracking of the skin (parakeratosis). While trace amounts of selenium are necessary for normal health, excessive amounts, which can be found in forages and grains in some regions, are toxic and may cause death. To furnish both calcium and phosphorus, grazing livestock may be allowed free access to such a mixture as 60 percent dicalcium phosphate and 40 percent common salt. Trace mineralized salt is used when copper or cobalt may be deficient. Livestock usually are given access to common salt separately, so they will not be forced to eat more of the other minerals than they require getting the amount of salt they need. Swine diets usually contain prescribed levels of calcium, phosphorus, salt, and essential trace minerals that may be deficient in the grains they are fed.

Vitamins

Known vitamins include the fat-soluble vitamins A, D, E, and K, and the water-soluble B group of thiamin, riboflavin, niacin, pantothenic acid, choline, biotin, folic acid, and vitamins B_6 and B_{12} and vitamin C.

Vitamin A, the one most apt to be lacking in livestock feeds, is required for growth, reproduction, milk production, and the maintenance of normal resistance to respiratory infections. All green-growing crops are rich in carotene, which animals can convert into vitamin A. Vitamin A supplement is added to animal diets to ensure a supply when livestock are not fed green forages and are not on good pasture.

Vitamin D enables animals to use calcium and phosphorus; a deficiency causes rickets in young growing animals. The ultraviolet rays of sunlight produce vitamin D from the provitamin in the skin. Field curing of hay develops vitamin D through the action of the sunlight on ergosterol in the hay crops. Certain fish oils are very rich in vitamin D. Livestock that are outdoors in the sunlight much of the time have a plentiful supply of vitamin D. Under winter conditions in cold regions, cattle, sheep, and horses ordinarily get ample amounts from the hay they are fed; pigs, poultry, and laboratory animals that are raised indoors will be deficient unless a supplement is added.

The vitamin B group is not important in the feeding of cattle, sheep, and other ruminants, because the bacteria in their rumen synthesize these vitamins. Very young calves, however, and poultry, swine, and other monogastric animals require the B vitamins in their diets. Of these, riboflavin, niacin, pantothenic acid, and vitamin B_{12} are most likely to be deficient in ordinary feeds; special supplements are needed by pigs, poultry, and laboratory animals. Choline may also be deficient in poultry feeds.

Vitamin E is necessary for normal hatching of eggs. It plays a role along with selenium in preventing muscle stiffness and paralysis (dystrophy) in lambs, calves, and chicks under certain conditions. Vitamin C, which prevents scurvy in humans and guinea pigs, can be synthesized in the bodies of most other animals and need not be supplied in their food. Vitamin K is synthesized by bacteria in the intestinal tract and can be absorbed, and, if livestock can ingest feces, a dietary supply is usually not important. Today many animals are raised without fecal contact, though, so vitamin K is often added to their diets as a safety factor.

Antibiotics and other Growth Stimulants

Antibiotics have been used in livestock diets since the early 1950s. They and other growth stimulants are non-nutritive substances added to animal feeds to treat diseases, to improve the efficiency of feed utilization and feed acceptance, or to improve the health or metabolism of the animal in some way. The use of antibiotics can be broadly divided into two categories, therapeutic and subtherapeutic, in which the distinction purely depends on the amount added to the feed. In therapeutic use, enough antibiotics are used to control bacterial infections within an individual or animal population; in subtherapeutic use, antibiotics are given in relatively low doses to enhance the performance (typically growth and feed efficiency) of the animals. The addition of subtherapeutic antibiotics to the diets of young pigs improves growth performance by 10 to 15 percent or more. Because the subtherapeutic use of antibiotics does not completely eradicate bacteria populations, over time the practice leads to antibiotic-resistant bacteria, which forces the search for new antibiotics for both livestock and humans. Some countries in the European Union have banned the subtherapeutic use of antibiotics, and there is growing pressure in the United States to ban the subtherapeutic use of penicillin and tetracyclines, the most important antibiotics for human use. The most commonly used antibiotics in feed are chlortetracycline, oxytetracycline, bacitracin, penicillin, and tylosin.

Other growth enhancers added to the feed do not have a disease-reducing affect but rather change the animal's metabolism. Most are related to hormones produced by the animals. These include lasalocid for cattle and sheep, melengestrol acetate and monensin for cattle, and ractopamine for swine. Several ear implants are approved for delivering hormones or other drugs to feedlot cattle in the United States. For example, these products typically increase daily gain by 10 to 15 percent and feed efficiency by 5 to 10 percent. Other countries, particularly those in the European Union, severely restrict or prohibit the use of implants in meat-producing animals because of opposition from some consumer groups.

Composition and Valuation of Feeds

The usual chemical analyses of feeds provide information on the amount of dry matter, protein (with its amino acid composition), fat, fibre, minerals, and vitamins contained in the feed. Various energy values (digestible, metabolizable, and net) of the feed, which depend on the species of animal being fed, are included in complete tables of feed composition.

Determination

Digestion and balance experiments measure the degree to which the various components of a feed are absorbed and retained by the animal body. Proteins are composed of varying quantities of amino acids, which are ultimately used by the animal to develop muscle and other body tissues. Microbes in the rumens of cattle and sheep can synthesize amino acids from the various nitrogen sources in their feed. In contrast, monogastrics such as pigs and poultry must have the proper amounts of the essential amino acids provided in their diet. Thus, ruminants typically have simple protein requirements, while monogastrics have amino acid requirements.

Protein and amino acid requirements are expressed as the amounts of digestible protein or amino acids needed for growth or other body functions, either as a percentage of the diet or as the total grams or units required per day. The amounts of energy needed are measured as digestible energy (DE), metabolizable energy (ME), net energy (NE), or total digestible nutrients (TDN). These values differ with species. The gross energy (GE) value of a feed is the amount of heat liberated when it is burned in a bomb calorimeter. The drawback of using this value is that a substance such as wood and corn may have a similar GE but vastly different nutritional values because some or all of it passes through the body without being digested. Furthermore, some of what is digested gets excreted in the urine as urea. Still more energy is lost from the heat of digestion and as gas produced by bacteria in the digestive tract. This loss is appreciably greater in ruminants than in pigs, chickens, and other monogastric animals. The work of eating, digesting, and metabolizing food may also be subtracted from the food energy, resulting in the NE value, or useful energy value of a feed that can be used for production (growth, reproduction, or milk) or for maintenance (basal metabolism, activity, keeping warm). The TDN value of a feed represents the sum of the digestible protein, digestible ether extract (fat) times 2.25, digestible nitrogen-free extract (carbohydrate), and digestible crude fibre. Its use as a measurement factor has declined in recent years because it considers neither the fermentation and heat losses during digestion and metabolism nor the fact that energy is utilized more efficiently for maintenance or milk production than for growth and fattening.

Optimization of Nutrient-cost Ratio

Feed costs vary widely from season to season; it is often possible for producers to realize substantial savings through wise selection of the feed ingredients used to formulate complete diets. It is much easier for large commercial feed companies with widespread operations to take advantage of regional variations in feed prices than it is for individual relatively small-scale livestock producers who must rely on locally produced feeds.

Least-cost formulation of feed mixtures makes it possible to use computers to select the correct amounts of competitively priced feed ingredients that will combine to fully satisfy the nutrient requirements of a specific type of animal at a particular stage of development. When used by a qualified nutritionist, computer programs can successfully formulate diets that yield maximum production at the lowest possible cost.

Basic Types of Feeds

Animal feeds are classified as follows: (1) concentrates, high in energy value, including fat, cereal grains and their by-products (barley, corn, oats, rye, wheat), high-protein oil meals or cakes (soybean, canola, cottonseed, peanut [groundnut]), and by-products from processing of sugar beets, sugarcane, animals, and fish, and (2) roughages, including pasture grasses, hays, silage, root crops, straw, and stover (cornstalks).

Concentrate Foods

Cereal Grains and their By-products

In the agricultural practices of North America and northern Europe, barley, corn, oats, rye, and sorghums are grown almost entirely as animal feed, although small quantities are processed for human consumption as well. These grains are fed whole or ground, either singly or mixed with high-protein oil meals or other by-products, minerals, and vitamins to form a complete feed for pigs and poultry or an adequate dietary supplement for ruminants and horses. The production of grains is seasonal because of temperature or moisture conditions or a combination of both. It is necessary to produce a full year's supply during the limited growing season. The grain is dried to 14 percent or less moisture to prevent sprouting or molding; the grain is then stored in containers or buildings where insects and rodents cannot destroy it. It is generally desirable to store more than a year's supply of the grains to be used as feed, because crop failures sometimes occur.

High-protein Meals

Vegetable seeds produced primarily as a source of oil for human food and industrial uses include soybeans, peanuts (groundnuts), flaxseed (linseed), canola, cottonseed, coconuts, oil palm, and sunflower seeds. After these seeds are processed to remove the oil, the residues, which may contain from 5 percent to less than 1 percent of fat and 20 to 50 percent of protein, are marketed as animal feeds. Cottonseed and peanuts have woody hulls or shells, which are generally removed before processing—if the hulls or shells are left intact, the resulting by-product is higher in fibre and appreciably lowers in protein and energy value. The protein quality of these meals for monogastrics varies greatly depending on the levels and availability of the amino acids present. Ruminants in general require only protein or nitrogen sources for the rumen microbes to synthesize amino acids.

These high-protein feeds supplement inexpensive roughages, cereal grains, and other low-protein feeds in order to furnish the protein and amino acids needed for efficient growth or production. The supplement chosen for a particular diet depends largely on the cost and availability of supply.

By-products of Sugar Beet and Sugarcane

From the sugar beet industry come beet tops, which are used on the farm either fresh or ensiled, and dried beet pulp and beet molasses, which are produced in sugar factories. Cane molasses is a residue from cane sugar manufacture. These are all palatable, high-quality sources of carbohydrates. Sugarcane bagasse (stalk residue) is fibrous, hard to digest, and of very low feed value. In Europe, beets and some other roots are grown as animal feed. Citrus molasses and dried citrus pulp, which are generally available at low cost as by-products of the citrus juice industry, are often used as high-quality feeds for cattle and sheep.

Other By-product Feeds

Large quantities of animal feed are by-products or residues from commercial processing of cereal grains for human consumption. The largest group of these by-product feeds comes from the milling of wheat, including wheat bran, wheat middlings, wheat germ meal, and wheat mill feed. In some areas, bakery wastes, such as stale and left-over bread, rolls, and various pastry products, are ground and used as filler or feed for pets and farm animals. Rice bran and rice hulls are obtained in similar fashion from the mills that polish rice for human food. Corn gluten feed, corn gluten meal, and hominy feed are produced as by-products from the manufacture of starch for industrial and food uses.

Brewers' grains, corn distillers' grains and solubles, and brewer's yeast are useful animal feeds and are collected from the dried residues of the fermentation industries that produce beer and distilled spirits. Waste products from pineapple-canning plants include pineapple bran or pulp and the ensiled leaves from the plant. By-products from the abattoirs and meatpacking plants that process animals into meat include such feeds as meat and bonemeal, tankage (animal residue left after rendering fat in a slaughter-house), meat scraps, blood meal, poultry waste, and feather meal. Various types and qualities of fish meals are produced by fish-processing plants. These animal by-products typically contain 50 percent or more high-quality protein and the mineral elements calcium and phosphorus. Steamed bonemeal is particularly high in these important minerals. Dried skim milk, dried whey, and dried buttermilk are feed by-products from the dairy industry.

Roughages

Pasture

Pasture grasses and legumes, both native and cultivated, are the most important single source of feed for ruminants such as cattle, horses, sheep, and goats. During the growing season they furnish most of the feed for these animals at a cost lower than for feeds that need to be harvested, processed, and transported. Hundreds of different grasses, legumes, bushes, and trees are acceptable as feeds for grazing animals. The nutritive value of the cultivated varieties has been studied, but information is incomplete for many of those that occur naturally.

Hay

Hay is produced by drying grasses or legumes when they approach the stage of maximum plant growth and before the seed develops. This stage has been shown to give maximum yields of digestible protein and carbohydrates per unit of land area. The moisture content is typically reduced below 18 percent in order to prevent molding, heating, and spoilage during storage. Legume hays, such as alfalfa and clovers, are high in protein, while the grasses (such as timothy and Sudan grass) are lower in protein and vary considerably depending on their stage of maturity and the amount of nitrogen fertilization applied to them. Stored hay is fed to animals when sufficient fresh pasture grass is not available.

Hay: Bales of hay.

Silage

Silage is made by packing immature plants in an airtight storage container and allowing fermentation to develop acetic and lactic acids, which preserve the moist feed. Storage may be in upright tower silos or in trenches in the ground. The initial moisture concentration of the forage should be between 50 and 70 percent, depending on the type of silage. Lower moisture levels can cause difficulty in obtaining sufficient packing to exclude air and may result in molding or other spoilage. Too high a moisture content causes nutrient losses by seepage and results in the production of excessively acidic, unpalatable silage. Ensiled forage can be stored for a longer period of time with lower loss of nutrients than dry hay. The nutritive value of silage depends on the type of forage ensiled and how successfully it has been cured. Corn, sorghums, grasses, and sometimes leguminous forages are used in making silage.

Making silage for cattle on a farm.

Root Crops

Root crops are used less extensively as animal feed than was true in the past, for economic reasons. Beets (mangels), rutabagas, cassava, turnips, and sometimes surplus potatoes are used as feed. Compared with other feeds, root crops are low in dry-matter content and protein; they mostly provide energy.

Straw and Hulls

Quantities of straw remaining after the harvesting of wheat, oats, barley, and rice crops are used as feed for cattle and other ruminants. The straws are low in protein and very high in fibre; digestibility is low. Straw is useful in maintaining mature animals when other feeds are in short supply, but it is too low in nutrition to be a satisfactory feed for extended periods unless supplemented with other feeds that supply the protein, digestible energy, and minerals needed for growth and production. Treatment of straw with alkali markedly increases the digestibility of the cellulose, augmenting its value as a source of energy for animals. Corncobs, cornstalks, cottonseed hulls, and rice hulls can also be used as sources of fibre in ruminant diets. Rice hulls are lower in value, while the others are similar to straw.

Animal Health

Monitoring animal health and preventing animal disease outbreaks is vital to the economy and safety of the country's food supply. Production of healthy livestock helps to ensure a safe food supply and keep consumer prices stable.

Importance of Animal Health

Animal disease outbreaks can cost the country millions of dollars due to animal slaughters, trade halts, and subsequent disease eradication efforts. For example, PRRSV, a virus that causes stillborn piglets, costs farmers an annual $600 million. Animal diseases with human health implications can adversely impact public health, global trade, and the stability of the agricultural segment of the economy. Healthy animals are more likely to result in:

- Safer food supplies.
- Higher farm productivity (including increased numbers of offspring).
- Reduced environmental impacts.
- Reduced use of antibiotics.
- Improved animal well-being.

3

Livestock Farming

The management and breeding of domestic or farm animals for obtaining meat and other products is known as livestock farming. There are various branches of livestock farming such as cattle farming, sheep farming, goat farming, horse farming, pig farming, rabbit farming, poultry farming, etc. The topics elaborated in this chapter will help in gaining a better perspective about these branches of livestock farming.

Cattle Farming

Cattle farming involve keeping bulls and cows for meat (beef) or milk (dairy) purposes. Cattle feed on grasses, legumes, roughage, etc. That is why they are called herbivores and they are also known as ruminants (because they have one stomach with four compartments).

Terminologies in Cattle Production

Some common terms in relation to cattle are:

- Bull: Adult male cattle.
- Cow: Adult female cattle.
- Calf: Newly born.
- Bull calf: Young male.
- Heifer: Young female.
- Bullock: Castrated male.
- Spayed: Castrated female.
- Calving: Act of parturition or giving birth.
- Serving or service: Act of mating.
- Gestation: Pregnancy and it is 279 to 287 days.

Is Commercial Cattle Farming Profitable?

Yes, a cattle farming is profitable. However, profitability depends on knowledge and proper management. This enterprise can be lucrative, based on the fact that management principles and fundamental knowledge have been known.

White Fulani Cattle.

Cattle farming is the management of bulls and cows for beef and/or milk production purpose. In commercial cattle farming, the goal of production is to make money through selling of live cattle or animal products (meat, milk, etc.). This enterprise is divided into three sub-production types, based on the purpose of production.

- Beef cattle farming: For meat (beef) production only.
- Dairy cattle farming: For milk production only.
- Dual-purpose farming: For both milk and meat production.

Just like any other farming business enterprise, you need to carry out a feasibility study before going to cattle farming business. Before anyone can start cattle business, he or she must be sure that feeds and herbage are sufficiently available for the cattle. Cattle are herbivores, i.e., they eat vegetables. Another thing about this business is that location and climate have a great influence on the performance of both bull (male cattle) and cow (female cattle). For example, cattle that have their native home in a temperate region may not survive or perform well in a tropical region. Therefore, before starting a cattle farming business, the following points must be noted.

Facts about Cattle Farming

- Cattle farming is highly profitable.
- It requires patience and could be a long term business if well-managed.
- It is a capital intensive farming business.
- It is a labor-intensive business.
- Cattle farming can be done under different systems which include feedlot system, cow-calf system, grass-fed system, grass-fed and grain-finished system.

- It requires a vast land size, property and adequate, clean water and water resources. This is because cattle drink a lot of water every day.
- A well-established and rich pasture is required.
- The farmer needs to have a good knowledge of genetics and breeding.
- Proper management is required in order to get the best out of the animals.
- Sheds and barns are very important. Trees should be on the farm as they will serve as shade when the ambient temperature is high.
- Fencing of the farm is very necessary.
- Services of experts such as veterinarians, nutritionists and breeders are vital.

Sheep Farming

Sheep farming is among the traditional business and occupations of the people of some countries around the world. Sheep have been rearing as a domestic animal from the ancient time. Usually sheep farming means 'rearing sheep commercially for the purpose of meat, milk and wool production'. Although sheep farming for commercial milk production is not a good decision.

Sheep are suitable for meat and wool production. If you have proper facilities, then you can raise sheep in both small and large scale. Commercial sheep farming business is very profitable and you will get your investment back within a very short period. Before starting sheep farming business, make a proper business plan and work according to the plan.

Sheep are raised mainly for their wool, milk, skins and manure production. Sheep meat is very tasty, nutritious and popular to all types of people throughout the world. Sheep farming business can be a great source of income and for eradicating poverty from the barren, desert, semiarid and mountainous areas. It is also a reliable income source for the people who are engaged with animal farming business.

- You don't need to have a huge capital for starting a sheep farm.
- You don't have to make an expensive house for your sheep.
- Sheep farming business require less labor than any other livestock farming business.
- Sheep give birth of kids frequently, so the size of your herd will be large within a short period.
- Sheep eat different kinds of plants, compared to other kinds of livestock animals. So you can use them for cleaning unwanted plants from your garden or field.

- Sheep hardly destroy trees than goats.
- Sheep can survive by consuming low quality grass and turn the feed into meat and wool.
- Sheep products such as wool, meat and milk are used for different purposes.
- They are very hardy animal, and can adopt themselves with almost all types of environment.
- Sheep require less space for living. Even you can raise sheep with your other livestock animals.
- By proper care and management, commercial sheep farming business can be a great source of earning and employment. Unemployed educated young can also make a good income and employment source through raising sheep commercially.

Importance of Sheep Farming in National Economic Growth

Commercial sheep farming can play a very important role to the national income of a country. Not only the sheep but also the goats have a great role in the national economic growth. By rearing sheep in small scale, the poor people can earn some extra income. Sheep products have a huge demand in the international market. So there is an opportunity to earn foreign currencies by exporting sheep products. That will directly increase national income.

Goat Farming

Goats are reared for milk and meat. Goat is a multi-functional animal and plays a significant role in the economy and nutrition of landless, small and marginal farmers in the country. Goat rearing is an enterprise, which has been practiced by a large section of population in rural areas. Goats can efficiently survive on available shrubs and trees in adverse harsh environment in low fertility lands where no other crop can be grown. Around the world, more people drink goat milk than cow milk. Also, more people eat chevon (goat meat) than beef. The World Health Organization says that more than 70% of the world's population has some allergy to cow milk. The allergic symptoms could be stomachaches, gas, skin rashes and ear infections. An allergy to goat milk is very rare. According to many historians, goats were the first animals to be domesticated. For thousands of years, they have been utilized for their milk, meat, hair, and skins all over the world.

Advantages of Rearing Goat

- The initial investment needed for Goat farming is low.
- Due to small body size and docile nature, housing requirements and management problems with goats are less.
- Goats are prolific breeders and achieve sexual maturity at the age of 10-12 months gestation period in goats is short and at the age of 16-17 months it starts giving milk. Twinning is very common and triplets and quadruplets are rare.
- In drought prone areas risk of goat farming is very much less as compared to other livestock species.
- Goats are ideal for mixed species grazing. The animal can thrive well on wide variety of thorny bushes, weeds, crop residues, and agricultural by-products unsuitable for human consumption.
- The goat meat is more lean (low cholesterol) and relatively good for people who prefer low energy diet especially in summer and sometimes goat meat (chevon) is preferred over mutton because of its "chewability".
- Goat milk is easy to digest than cow milk because of small fat globules and is naturally homogenized. Goat milk is said to play a role in improving appetite and digestive efficiency. Goat milk is non-allergic as compared to cow milk and it has anti-fungal and anti-bacterial properties and can be used for treating uro-genital diseases of fungal origin.
- Goats are 2.5 times more economical than sheep on free range grazing under semi-arid conditions.
- Goat creates employment to the rural poor besides effectively utilizing unpaid family labour. There is ample scope for establishing cottage industries based on goat meat and milk products and value addition to skin and fibre.

Selection of Goat

Malabari (Tellichery), Attappady, Sannen x Malabari cross-breeds are available in the state. When buying an adult goat, be sure to check its milk production. Milk yield per day assessed by recording two consecutive milking, should be more than 0.5 kg (including milk sucked by kids). When selecting young goats, the dam's production may be checked. A one year old she-goat should weigh about 20 kg. Doeling at 6 months should weigh not less than 10 kg. The doeling must also be free from physical defects. Selection of does should be based on their previous 120 days' milk production record. Those, which have kidded at least by 2 years of age, should be preferred.

Goat Production in the World

Figure shows the global distribution of goat production between 1994 and 2014. In the last two decades, we can see that the geographic areas with the largest goat production are located in the global African continent, in a range from Southeastern Europe (Greece and Turkey) to South Asia (India and China), passing through the Arabian Peninsula and, in the American Continent, where the highlight goes to Mexico, Brazil, and Argentina. The production of small ruminants is largely rooted in some world regions, for both historical and religious reasons. This is the case in all Muslim countries where pork is not a food option.

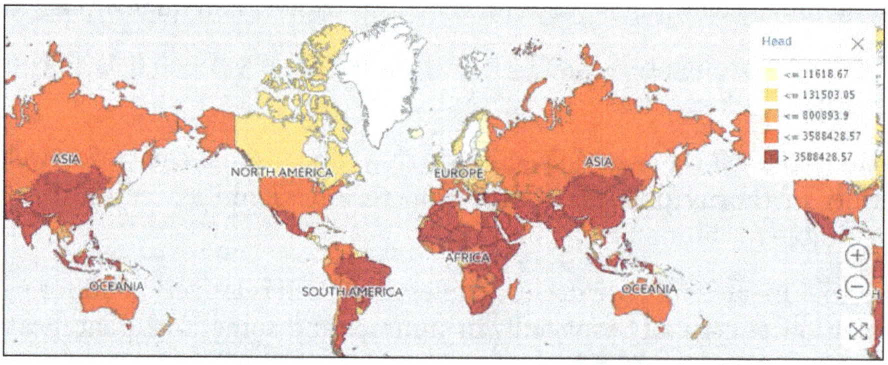

Global distribution of goats by country in number of heads.

Soil and climatic matters greatly condition the animal species produced, benefiting goat production in poorer areas in terms of forage capacity and enhancing this species in areas with steep slopes and mountainous rocky areas. Sometimes goat production is the only possible, due to the excellent adaptability of these animals to the difficult mountain conditions. On the other hand, forest grazing is an excellent productive alternative, associating in an integrated way the forest and animal productions. It helps to control unwanted weed growth with a consequent strong reduction of fire risk and, on the other hand, is often a valuable complement to the farmer or forest producer income, while it also helps to fertilize forest soils with their waste. To confirm this evidence, we can also verify in figure that goat production reaches high levels in the most important forest and mountain areas.

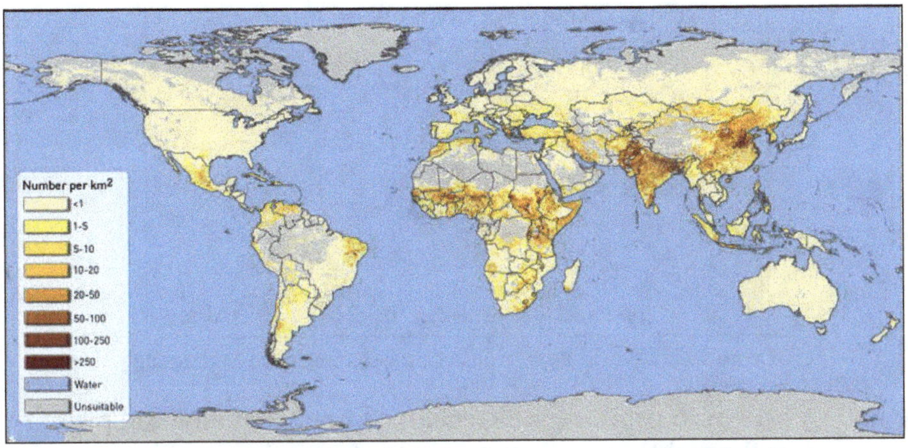

Worldwide distribution of caprine by country, in animal density.

In figure below, we can see similar results to those in Figure, but here we can see the distribution by animal density, in number of heads per km². Here we find the highest values in Mexico, northeastern Brazil, sub-Saharan tropical strip, India, and Eastern China.

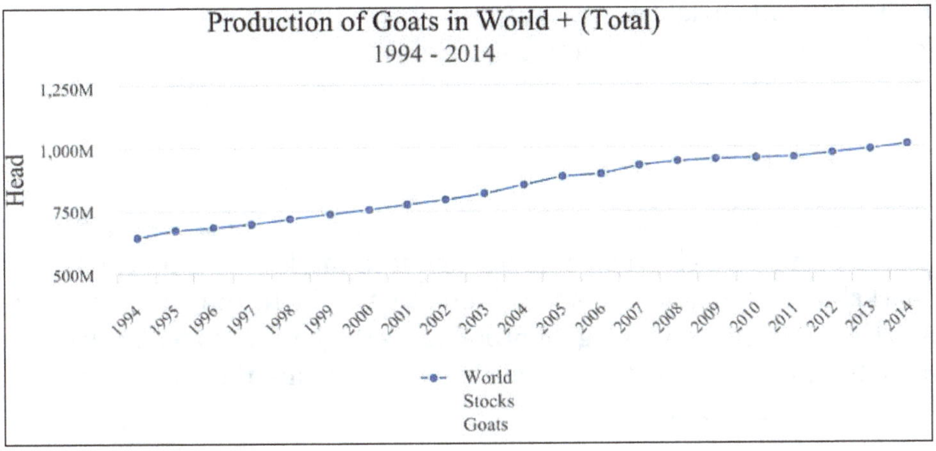

Evolution of world production of goats between 1994 and 2014.

Figure above shows that, only in the referred 20 years, the world goat population (from which there is a register) go from slightly over 600 million head in 1994 to over 1 billion in 2014, 67% rise in just two decades, much higher than the world population growth in the same period, which was about 28.5%.

The Asian continent clearly leads the world goat production between 1994 and 2014, a percentage well above the sum of the non-Asian world goat production. Then, we can find the African continent with about a third of the total, and finally, with very low percentage values, we have America, Europe, and Oceania, in this order, which make up, in total, about 7% of global goat production. These figures complement and confirm those reported in Figure, where sub-Saharan Africa, China, and India showed their worldwide leadership in the production of this species.

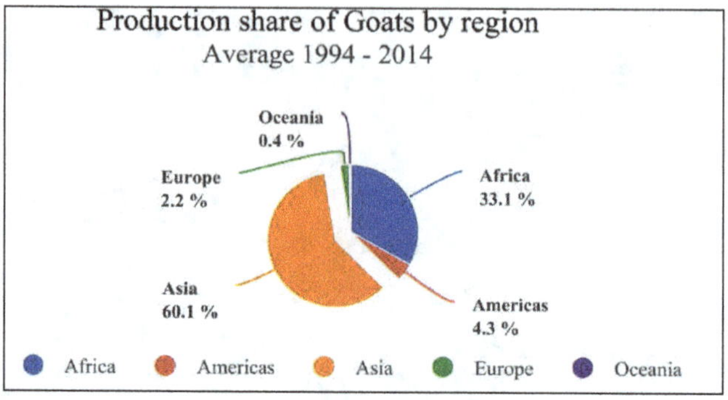

Evolution of world goat production between 1994 and 2014.

Goats are mainly grown to produce milk, meat, or fiber. Usually, the primary production is meat and milk, and there is little global information on fiber production. The potential of goats for the sustainable supply of milk and meat for human consumption is unquestionable, and their contribution to improve the nutrition of rural populations is likely to increase. At the same time, consumption of goat cheese should also increase in developed countries. This is attributed to the image of goat cheese as a product obtained under natural conditions when compared to cow's milk and its products, obtained from more industrialized farms. Government programs to support goat rearing should focus on research and training on genetic improvement, farm management, infectious disease control, product collection and marketing markets.

Milk Production

Unfortunately and unfairly, the dairy goat is still considered in some regions as the cow of the poor. In proportional terms to its size and feed intake, the milk yield of the goat is often remarkable. The goat has a reduced food intake, its feeding is cheap, it is a small animal and easy to keep, and each goat produces enough milk to feed an average family. Comparatively, and due to its size, the cow has very high maintenance costs and is more difficult to maintain at home.

Dairy goats produce about 15.2 million tonnes of milk, representing about 2% of the world's total milk produced by livestock species. Developing countries, particularly in Asia and Africa, account for about 83% of the total value produced. In Europe, dairy goat farming is the main reason for goat production, where only 3% of the world goat population produces about 15% of the global goat milk, mostly for cheese production. Spain, France, and Greece are the main European producers of goat's milk. In France, the development of research and development in the dairy goat sector has been intensified with programs organized for the selection, processing, and commercialization of goat's milk, mainly produced from the Saanen and Alpine breeds. In France, it is where the highest production figures per animal are achieved, whereas Iran has the lowest values.

China has the largest number of goats in the world, but they are bred mainly for meat production. Milk production by milk breed occupies the third place, behind France and Spain. According to China's official statistics, there are between 1.4 and 5.8 million dairy goats producing about 0.3 million tons of milk. The statistics on these goats can be questioned because they are not based on actual censuses, and knowledge of the number of dairy goats is even more difficult.

At the global level, interest in dairy goats has steadily increased, and in less than two decades, milk production has grown from around 10 million tons in 1990 to about 15.2 million tons in 2008. In developing countries, goat milk production has continued to grow partly due to a trend toward self-sufficiency of rural populations, where goat milk is the nutrition basis of millions of people. In these regions, milk is largely consumed raw but can also be processed into a wide variety of products for better preservation and commercialization. We can say that the global marketing of goat's milk is still at a very early stage, and so far, there has been no large-scale effort to professionalize this industry, exception for some developed countries. The marketing rate of goat's milk at the global level does not exceed 5%.

There are still many obstacles to achieve the professional level of production and sale of goat's milk. Access to productive data to enable the development of breeding programs is very scarce, and on the other hand, reproductive and productive seasonality, which is very marked in some breeds, prevents a constant supply of milk demanded by markets. In sum, there are consumers for goat dairy products and there is enough production to meet this demand, but there is a lack of bridges between both sides in order to enhance this market in terms of an efficient marketing of the product toward profitable industry.

Meat Production

Goat meat is widely consumed in developing countries. Of the approximately 280 million total tonnes of meat consumed per year, goat meat accounts for only 2% of this total, about 4.9 million tonnes. Developing countries produced approximately 97% of this amount, reflecting the great importance of goat meat to feed these populations. China leads the world production of goat meat, representing 38% of the total. As we could see in milk production, also in goat meat production the Asian and African continents are leading. As in milk production, the increase was also notable in the same 18 years of comparative production and raised from the 2.65 million tonnes of goat meat produced in 1990 to 4.93 million in 2008. Most of this meat is not commercialized but is produced and consumed locally.

The goat meat business is not already become profitable on an industrial basis. It is not enough to know the proper management techniques, but nowadays it is also essential to control the market and the economic limiting factors for farming success, as well as the financial management and marketing techniques. In addition, there are an increasing number of challenges associated with meat production, including knowledge of regional preferences and consumer education for consumption. In the production and processing sector, producers have to undergo training programs, and slaughter and processing facilities have to be modernized. Finally, there has to be made a strong commitment to organized programs of genetic improvement, market analysis, and marketing channels. The increase in goat meat consumption has been occurring in some parts of the world due to the migrations of people who prefer goat meat to the developed countries. This recent fact has increased demand for goat meat in these areas.

Characteristics of Goat Production Systems

Goat production systems, like other production systems, are not just a combination of crops and animals with the aim of achieving short-term results. They represent a set of interacting elements, managed by the farmer, according to their objectives. The farmer will define the objectives influenced by the social environment in which he is inserted, his degree of technical knowledge, and the available production factors.

The general and strategic management approaches are applicable to agricultural enterprises and those who apply them are more successful. The application of strategic management concepts helps to design the way the farmer will achieve the objectives, and it is convenient to know the strengths and weaknesses, the opportunities and the threats.

A caprine production system has production subsystems, whose performance influences the overall results, for example, the forage system; if there are nutrient deficiencies in the soil, its production is lower, and then, the animals may not have adequate feed to their needs. The increase in the meat and milk ruminant production has been much greater than the increase in the pasture area, mainly because this increase was due to the increase of mixed or landless production systems (intensive systems) than to the pastoral or extensive systems.

Increasing livestock production to provide food to the growing human population has increased the potential to cause environmental problems, and the balance between the environment and livestock production is now a concern. It is important to avoid an increase of grazing land and arable land for crops production to feed the animals, with the consequent reduction of areas of natural vegetation, soil and air contamination. Overgrazing also leads to several bad consequences like high amounts of animal excretions and overuse of natural pastures with low productivity and with high risk of soil degradation, especially in the arid and semiarid regions of the tropics and subtopics.

The terms equilibrium, disequilibrium, and nonequilibrium are commonly used and can be understood by equilibrium when the animal population and forage resources are under stable climatic conditions. Otherwise, climatic variations disturb the system, causing disequilibrium, which differs from nonequilibrium, represented by modifications in vegetation due to changes in the proportion in the plant species or where the dynamics of the animal population is disassociated with key factors that determine their survival, and their survival is maintained through feeding supplementation.

Goats are the best adapted ruminants to highly anti-nutrifibrous low-protein forages, often in conditions of poor availability of water. Goats are opportunistic feeders, the time they spend on grazing species depends generally on the relative frequency of encounters, but this relationship depends on species of vegetation and habitat visited.

Compared to other domestic ruminants, namely bovine and sheep, raised in regions of poor agricultural resources, goats' advantage is clear. These animals combine the advantage of being able to feed on a variety of low-quality fodder and shrubs, they also manage to walk long distances, with short breeding intervals with high reproductive rates, they provide high rates of investment return and, consequently, low investment risk. Goats also have high energetic efficiency in milk production, excellent utilization of marginal lands, a very strong flocking instinct and a docile behavior, which enables herding by children and elders.

Farmers tend to have mixed herds of sheep and goats as a strategy to maximize the use of environmental resources. Small ruminants in many traditional systems in the Mediterranean basin are the main source of red meat for human consumption, while in Northern Europe, in addition to meat, the milk, wool, and skin were also the main products. However, with development for market agriculture, meat has become the main product in the sheep sector, while in the goat sector milk has been, and still is, the main product while meat is secondary in most cases followed by skins and hair. There are, however, local variations where a particular product can be valued such as sheep's milk for the production of traditional cheeses and the production of goats in the southern Mediterranean. These products are mainly consumed regionally, constituting market niches with low international visibility.

Often, small ruminants, especially goats, are extensively produced using the poorest land, shrubland, and forest areas where other species cannot survive. This helps to fix the rural population, reduces the risk of depopulation of marginal or less-favored areas, and contributes to the maintenance of good agro-environmental practices and landscape preservation.

Models of Goat Production Systems

Extensive Goat Production Systems

Extensive systems are characterized by large areas to feed the animals, with a low animal density. It uses soils of poor agricultural ability, located in mountainous areas with large rainfall or in areas of low rainfall, sometimes with extreme temperatures. There is a use of natural resources, made by autochthones goat breeds that are perfectly adapted to the environment, very rustic, but with low productivity.

The goat production in this extensive system uses family labor, often as a second source of income for families. The goat milk is obtained for family consumption or for cheese to sale. However, the sale of the kid goat is the main reason for this production. In the

Southern Mediterranean, this product is very appreciated and valued, mainly in the Easter and Christmas seasons. Usually, the number of animals produced is low and they are usually raised under climate adversities, where probably there are no shelters, no food supplementation and also often these animals have hygienic sanitary problems, which lead to poor economic results.

It turns out that the use of natural pastures can be done in two ways, a mobile grazing system and a sedentary grazing system. The first is characterized by annual or seasonal movements of the animals with the shepherd to new places in search for feed, and while sedentary grazing, the animals are driven freely to pastures near the farm, usually keeping the animals at night in the stable.

In the mountains, they can practice transhumance, which consists of seasonal movement of animals regularly between two or more areas of seasonal pasture, through established paths (called "canadas" in Portugal and Spain), conducted by shepherds in the summer to the mountain to take advantage of the still fresh pastures and return to the valley in early autumn. This modality is currently disappearing due to sanitary issues and also because of the greater easiness to feed the animals through the purchase of commercial feeds. Although the extensive production is not very productive, it is of great importance regarding the maintenance of the rural landscape and with the aim of the biomass management that avoids the occurrence of forest fires, and where the goats are well adapted to take advantage of these feed resources.

The use of goats in extensive systems can be valued by the quality products. In Europe, many of them have protected designation of origin (PDO) and the protected geographical indication (PGI) which are certified to attest the traditions and specific product qualities, strongly linked to a certain region. The awareness of society about the damaging effects of intensive livestock systems has changed the methods and aims of researchers and even research centers, trying to focus on improving system sustainability instead of increasing productivity.

Intensive Goat Production Systems

Intensification is often associated with a decrease in grazing dependence and an increase in the use of concentrated feeds, mainly cereals, to supplement natural feeds. At the same time, improved and balanced feeding practices together with improved breeds in ruminant systems enabled more efficient feed ratio conversion to meat and milk production rather than to maintenance of the animals.

The intensive system implies a high density or animal concentration per area unit, under reproductive and sanitary control, and the feeding process includes advanced technologies. Sometimes, some farms seasonally require higher feed and labor resources such as through the calving season and milk production, so the supplementing of animals with concentrated feed may be needed at this point, but the remaining year is mainly grazing.

In order to meet the feed requirements of animals in an intensive system, pastures must have high dry matter yields per hectare, good growth throughout the year, in both regions with regular rainfall or holdings with irrigation systems, without extremes of heat or cold temperatures. It is important that the breeds used in the intensive system have a high fertility and growth rates than the adaptability to the environment, that is, the rusticity. The equilibrium with the environment can easily be compromised by the insertion of exotic breeds, as well as with exotic or genetically modified plant species. The intensive use of natural resources may lead to their depletion and increase the environment pollution, having serious social consequences.

Goats are well adapted to harsh environments but can also be used in intensive systems with permanent housing, as is happen with many farms for milk production. Here too, mainly because these farms are usually in regions with weak resources, they use less labor but with higher qualifications, with the danger of triggering a process with social negative consequences, namely with soil erosion, high risks of fire and depopulation. Also industrial production can replace the artisanal production, losing ancestral traditions and biodiversity.

Intensive goat farms have higher costs for the installation and maintenance of production than extensive farms, but their choice should allow for a balance between environmental, economic, and social factors, since only then will it be sustainable.

Table: Main differences between intensive and extensive production systems.

Extensive system	Intensive system
Eco-agrarian conditions	
Dependent on the climate (environmental) factors	Independent of the climate factors
Scarce resources	Constant forage production
Intermittent feed cycles	Continuous feed cycles
Parasitic problems	Good sanitary condition
Use of marginal land and nonagricultural resources	Use of technology
Low animal density	High animal density
Animal biotypes	
Autochthonous breeds	Improved breeds
Low productions	High productions
Multi-production exploration	Animals with high energy efficiency
Great ability of adaptation	Increased sensitivity to diseases
Lower fecundity and fertility	Increased fecundity and fertility
Environment	
Natural environment	Artificial environment
Long productive cycles	Short productive cycles
Traditional system	Industrial system

Horse Farming

There are many things that set horses apart from other livestock:

- To start, there's the human-horse bond that many people develop. More people are apt to make their horses their pets than with, say, cow. This could be because of the closer interaction that people have with their horses. If a horse is kept for riding, driving or other sports, their caretakers interact with them just about every day, learn their personalities, and pay attention to their physical well-being—from muscle soreness to scrapes and bruises. It's easier to become bonded to an animal when you have this type of relationship versus simply feeding and opening a gate for it.
- Physiologically, horses differ from other animals, too. In terms of digestion, cattle, sheep and goats are ruminants; pigs, dogs and humans are nonruminants; and horses are somewhere in between, classified as nonruminant herbivores. Horses' digestive systems feature enzymatic digestion that occurs in the foregut, plus fermentive microbial digestion in the hind gut, making them able to breakdown forages.
- This single-stomach digestion system makes horses more susceptible to digestive issues. Colic is the name for general abdominal discomfort, and it can be caused by any number of things ranging from having eaten something moldy or toxic to having a heavy parasite load. Unlike ruminants, horses cannot regurgitate or burp, so they're often unable to relieve the pressure built up from intestinal issues. For all of these reasons, special care needs to be given to the pasture management, nutrition plan and water intake of horses.
- Horses' physical structure is different than other livestock, too, because we ask them to use their bodies for their everyday work. Unlike a pig bound for market, whose job it is to gain weight and wallow in the mud, horses are athletes that have a lot of moving parts that are susceptible to sprain, strain and injury.

Farm Management

A stable-management job can vary based on the type of farm you want to operate. For example, if you have a breeding farm, you'll have all of the regular farm-management duties plus stallion management, care of foals and assisting in foaling. If you manage a small-acreage horse farm where boarders take on "self-care"—daily care of their own horses—you might primarily be responsible for farm and facility maintenance but little feeding or stall cleaning. There are also performance-horse facilities that have indoor riding arenas to be cared for, clients paying thousands of dollars each month for their horses' care and multiple horse trainers working in the barn, so stable management here will require a lot of people skills in addition to basic farm-management know-how.

Create Productive Pasture

Horse farms—and small-acreage horse farms, in particular—easily become overpopulated with horses, making it hard to maintain pastures. In nature, horses spend 13 to 18 hours per day grazing, and this can quickly knock out a quality stand of forage. Horses' large, unforgiving hooves can quickly turn a pasture into mud, making wet-season turnout a problem. Their grazing habits involve selective eating, meaning they eat what they like first and then will eat other, less-desirable plants. They also select a few areas as bathroom areas, depositing their manure in the same general area and not grazing that area, which can waste valuable grasses.

Advantages to keeping a pasture in good shape include the ability for horses to get much of their nutritional need from pasture—sometimes all, depending on the horse, the weather, and the forage quality and density. From a land-stewardship perspective, productive pasture is good for the environment, acting as a filter for nutrients, providing a habitat for microbes and insects, and preventing erosion.

Minimize Mud

Grass doesn't grow in mud, of course, so if an area is turned to mud, it has to be rehabilitated, replanted and given time to rest before it can be used as productive pasture again. The moisture in mud also softens horses' skin and breaks down the skin's defenses, allowing bacteria to easily enter. Thrush and scratches are common mud-related issues seen in horses. The uneven surface and suction action of mud can injure horses and pull their shoes, too. Tools to manage mud differ based on facilities available, your climate and your budget. These might include adding gutters to the barn, having a sacrifice lot, turning out horses in the indoor arena when outdoor grazing isn't an option, and complete pasture renovation.

Monitor Horse Health

Healthy horses are, of course, the main goal for a horse-farm manager. You may be the person in charge of scheduling veterinarian and farrier visits and ensuring horses are on a regular farrier, vaccination and deworming schedule. When horses get sick or injured, it's often the farm manager who notices and cares for the animal until the owner and veterinary help arrive. Basic equine first-aid skills are a must for anyone involved in equine management.

Manage Manure

If horses are good at one thing, it's making manure—approximately 50 pounds per day. Manure management should be taken seriously, as the nutrients leeching from manure can contaminate waterways, manure attracts flies, and no one wants to smell your horse farm's mountain of manure. Your manure-management options depend on your horse farm's size, location and overall goals. Horse manure is a valuable crop amendment and could be composted or sold as-is to local farmers or used on your gardens. Some areas with large horse populations have manure-removal services, too.

Pig Farming

Pig farming is adapted to both diversified and intensified agriculture. Pigs convert inedible feeds, forages, and certain grain by-products obtained from mills, meat by products, damaged feeds and garbage into valuable nutrition's mean. The faeces of pigs are useful in maintaining soil fertility.

The pig grows fast and is a prolific breeder, farrowing 10-12 piglets at a time. It is capable of producing two litters per year under good management conditions. The carcase yield is high: 65-80% of live weight. With a small investment on building and equipment, proper feeding and sound disease control program the farmer can profitably utilize his time and labour in this subsidiary occupation.

Selection of Breeds

- For commercial pig farming upgraded/crossbred or exotic stock in good health should be selected.
- While selecting a gilt or sow primary aim should be to secure a female that will produce large survivable litter and which can attain marketable weight at an age of six months or less.
- Animals those are ready to be bred should be selected.
- Identify the newly purchased animal by giving suitable identification mark (ear notching or tattooing).
- Vaccinate the animals against diseases.
- Keep the animal under observations for a period of about two weeks and then mix with the other animals.
- Purchase animals in two batches at the interval of three months.
- Follow judicious culling and replacement of animals in a herd.
- Cull the old animals after 10-12 farrowings.

Breeding Boars

Selection

The following are the important criteria for selection of boars for breeding purposes:

- They must be offspring of better producing gilts and sows.
- They must belong to a litter with size and weight of not less than 8 and 72 kg respectively at weaning.
- Boars satisfying above stipulations are subjected to preliminary selection at 5 months of age provided they have a body weight of at least 60 kg.
- Final selection of boars should be undertaken at 7 months provided it has attained a body weight of at least 90 kg.
- Must be free from physical defects.
- Breeding males and females should be selected from different litters and boars should be changed periodically – preferably once in two years – for avoiding inbreeding.

Modern and well-established scientific principles, practices and skills should be used to obtain maximum economic benefits from pig farming.

Management of Boars

- Although boars reach sexual maturity at 7 months of age, it is generally recommended that they must be at least 10-12 months (100 kg body weight) before being put to regular use.
- Boars should be fed after service rather than before.
- Boar: Sow ratio may be 1:25.

- Outdoor exercise is good for keeping the boar thrifty and virile. He should neither be over fat nor run down. The boar may be handled firmly but gently, keeping in mind that he can be dangerous.
- Each boar should be housed separately.
- Number of services for boars: 2 - 4 per week depending on the age of boar.
- Boars should be tested for brucellosis and leptospirosis periodically.

Breeding Females

Selection

- Gilts: Breeding pigs are selected from mother pigs having good mothering ability as shown by large litter size and weight at weaning. Gilts should have 12-14 evenly spaced sound teats, good growth rate and femininity.

Gilt

- Weaned sows: Sows are usually weaned at 8 weeks after farrowing. The weaned sows come into heat in 3-10 days after weaning. They may be bred on the first post-weaning heat itself since this period is most fertile. She may respond to nutritious diet and pick up fast if the condition is poor.
- Pregnant sows and gilts: In addition to having proper nutrition, the pregnant animals should be provided with regular exercise. Forced exercise may be brought about by feeding some distance away from the house or inducing the animal for moderate walk. Sows of the same size and condition can be run together. The bred and gilts may be housed separately. Do not house boars along with pregnant stock. Avoid overcrowding, mixing of new and old stock, slippery ground and overexciting of pregnant pigs.

The advantages of the pig farming are:

- The pig has got highest feed conversion efficiency i.e. they produce more live weight gain from a given weight of feed than any other class of meat producing animals except broilers.

- The pig can utilize wide variety of feed stuffs viz. grains, forages, damaged feeds and garbage and convert them into valuable nutritious meat. However, feeding of damaged grains, garbage and other unbalanced rations may result in lower feed efficiency.
- They are prolific with shorter generation interval. A sow can be bred as early as 8-9 months of age and can farrow twice in a year. They produce 6-12 piglets in each farrowing.
- Pig farming requires small investment on buildings and equipments.
- Pigs are known for their meat yield, which in terms of dressing percentage ranges from 65 - 80 in comparison to other livestock species whose dressing yields may not exceed 65%.
- Pork is most nutritious with high fat and low water content and has got better energy value than that of other meats. It is rich in vitamins like thiamin, Niacin and riboflavin.
- Pigs manure is widely used as fertilizer for agriculture farms and fish ponds.
- Pigs store fat rapidly for which there is an increasing demand from poultry feed, soap, paints and other chemical industries.
- Pig farming provides quick returns since the marketable weight of fatteners can be achieved with in a period of 6-8 months.
- There is good demand from domestic as well as export market for pig products such as pork, bacon, ham, sausages, lard etc.

The swine production generates slurries nutrients rich, which could be revaluated in cereal crops used for its food and energy generation (biogas) for use on the farm. However, the revaluation requires knowing their physical–chemical and biological characteristics, which allow giving an adequate transformation (treatment). On the one hand, swine production and consumption market reveal the superiority of emergent countries on meat/cereal (feed) production and swine meat consumption (concentrated population). The food composition and growth phase will influence the swine slurries composition, which is rich in organic matter, macronutrients (N, P) and micronutrients (Cu and Zn). These characteristics will generate odors (organic matter, macronutrients) and ecotoxicology effects (macro/ micronutrients) if they are not treated.

Moreover, the swine slurries treatment allows revaluated them in agriculture and obtaining energy. Anaerobic technologies (anaerobic lagoon, mixed complete reactors, UASB, among others) are the most used/cost-effective to organic matter removal from swine slurries, obtaining from 0.28 to 0.83 m^3 biogas/kg organic matter. Meanwhile, passive technologies (constructed wetlands) are the most used technologies to nutrients and metals removal. Treated swine slurries from constructed wetlands have agronomic properties. Therefore, the nutrients cycle within swine production would favor concepts of revaluation in origin.

The economic and demographic growths are the factors that are activating the current world meat production. Thus, developing countries where 75% of the population is located also concentrate 52% (41.5 million ton/year) of swine production. Moreover, in the last 40 years the growth of emergent economies are concentrating both meat and cereal production in one place.

On the other hand, demand is the main trigger factor of the current "livestock revolution". Thus, the meat consumption acquires connotation different depending on the consumer type. In developed countries, higher purchasing power and demanding lifestyles have promoted a higher swine meat demand (25 kg/per capita year). Meanwhile, developing countries consume only 8 kg/per capita year, condition related mainly with subsistence habits. However, developing countries (75% worldwide population) concentrates 64% (43.3 million ton/year) of the swine meat total consumption; while that, developed countries concentrate 36% (34.4 million ton/year) remaining. However, emergency economies (Brazil and China) are increasing about 21% the swine meat consumption per capita of their population, concentrating swine meat consumption in these countries.

Swine Slurry

The swine slurry generation is related to the growth phase and its water/food requirements. Thus, lactation and reproduction phases have a great water supply (12.2–41.1 L water/animal d), mainly related to hydration improvement or fertility. Meanwhile, fattening and weaning have a greater food intake (1.9 kg food/ animal d), exclusively given by increased weight or age.

The swine diet is based on proteins, carbohydrates and starch concentrates, which exceed 50% of food total composition. Indeed, fattening tends to prioritize the protein intake (>30%); while, reproduction and weaning consume mainly fiber (10%) to avoid overweight. In both cases, the low food digestibility could generate nitrogen excretion (<30%) and phosphorus (<10%). This factor can be observed in the swine conversion index, which can vary from 2.0 to 4.3 and 0.9 to 2.3 kg feed/kg weight gained for fattening and previous phases, respectively. Indeed, digestibility can vary depending on the food and animal type, being slightly lowers in the final growth phases (fattening ~ 25%) than initial phases (weaning >50%).

Finally, the water/food ratios allow indirectly evaluating the swine slurry quantity and its composition. Thereby, initial phases (weaning) generate the water/feed ratios from 4.4 to 5.1 L water/kg food; while, final phases (fattening) reaches values from 1.6 to 2.5 L water/kg food. This indicator is indirectly related with nitrogen concentrations from urine/feces values. Indeed, greater urine/feces values are obtained by fattening

(6.6–9.1) than weaning (5.5–6.3). At this point, it's important to note that within the swine excreta around 75% of the nitrogen is in urine, so a higher water intake will increase the nitrogen generation in the slurries.

Table: Productive indicators within swine production.

Feed type	Conversion index kg feed/kg weight gained	Water/food ratio L water/kg feed
Pellet	2.13–3.33	–
With/out phytase	2.00–4.32	–
Wheat and soybeans (12–20%)	2.22–2.35	–
Intensive/organic	2.70–3.20	–
Ad libitum	2.95–3.05	1.99–2.60
Wean/mat/fat food	0.98–2.14	2.00–20.00

Generation and Physicochemical Characteristics

The physicochemical composition from swine slurries will be influenced by the low digestibility of nitrogen (<33%), phosphorus (<32%) and micronutrients (<3%), as well as the animal growth phase. Initial phases (maternity and lactation) produce more slurries (10.0–41.1 L/animal d) than weaning-fattening (3.5–9.10 L/animal d). This could be related with dilution by more water intake. Table summarizes the physicochemical characteristics from swine slurries according to the growth phase. The main differences are related to a higher content of organic matter (1.8–2.4 times), nitrogen (1.3–2.8 times) and phosphorus (2.1–4.6 times) from fattening with respect to maternity and weaning. This condition is related to a higher food intake (1.7 times), which is made up of carbohydrates and proteins.

Table: Physicochemical characteristics from swine slurries.

			Growth phase	
Parameter	Unit	Maternity	Weaning	Fattening
Slurry	L/animal d	10.0–16.0	23.5–41.1	3.5–9.1
pH		7.5	6.9	7.2–8.4
EC	mS/cm	12.8–15.5	14.2	15.3–25.3
TS	%	1.7	2.7	3.2
BOD_5	g/L	9.0	25.0	16.6–21.6
COD	g/L	24.0	65.2	45.3–57.7
TN	g/L	1.8–2.3	2.3	2.4–4.6
$N\text{-}NH_4^+$	g/L	1.4–1.8	1.5	2.0–3.1
TP	g/L	1.4	0.6	0.8–2.8
K	g/L	2.2	1.8	1.9–3.8

Cl^-	g/L	3.6	2.3	5.1
Cu	mg/L	11.0	55.0	15.9–80.0
Zn	mg/L	75.0	533.0	40.7–191.0

Meanwhile, micronutrients, such as zinc are excreted (2.7–13.0 times) in the initial phases (weaning). In a different proportion, copper is excreted by swine slurries, increasing during fattening (1.5–7.2 times). Differences in the presence of copper and zinc excreted is related to the use of both metals as growth promoters and specifically copper that is used additionally for therapeutic purposes (fighting diarrhea) in the most vulnerable swine population (weaning). Chloride, ammonium and potassium salts may also be present within swine slurries. The feeding has influenced a greater salts excretion (1.4–2.3 times) during weaning fattening. These characteristics are related to the fact that the salts in the swine feeding, favor the liquids retention increasing weight during the growth.

On the other hand, raw swine slurries show agronomic properties (N:P: K between 1.1:0.6:1.0 and 1.3:0.4:1.0). Indeed, the nutritional requirements of cereal crops (e.g. corn, wheat) can reach values of N:P: K between 1.2:0.2:1.0 and 1.6:0.3:1.0. However, the presence of pathogenic organisms and micro contaminating (metals, antibiotics) may limit their use prior to treatment.

Ecotoxicological Characteristics

Swine slurries have ecotoxicological characteristics depending on the bioindicator used. Tables below shows the ecotoxicological characteristics (swine slurries and composition) on terrestrial and aquatic bio-indicators. The ecotoxicological studies have been carried out mainly in Daphnia magna, establishing higher acute toxicity at low concentrations (48 h-LC_{50}, 1.8–3.3%). Meanwhile, chronic ecotoxicity has been observed on Lepidium sativum L. (growth inhibited at concentrations from 3 to 10%, v/v). However, Eisenia foetida growth (6000–20,000 worms/m^2) has also been reported at concentrations less than 25%.

Table: Ecotoxicological characteristics on aquatic organisms.

Organism	Test	Compound	Concentration (mg/L o % *)
Swine slurries			
Daphnia magna	48–24 h-LC_{50}	Raw swine slurry	1.8–5.0
Nutrients			
Moina macrocopa	24 h-LC_{50}	NH_4^+	231–492
Penaeus semisulcatus	96 h-LC_{50}	NH_4^+	11.4–55.9
Ceriodaphnia dubia	48 h-LC_{50}	NH_4^+	20
Daphnia carinata	24–48 h-LC_{50}	NH_4^+	2.2–2.8

Monia australiensis	24–48 h-LC_{50}	NH_4^+	7.5–8.5
Daphnia magna	48 h-LC_{50}	NH_3	2.9–6.9
Metals			
Daphnia magna	48 h-LC_{50}	Zn	0.05
Daphnia magna	48 h-LC_{50}	Cu	0.56
Daphnia magna	LOEC 14d	Cu, Zn	0.12
Danio rerio	96 h-LC_{50}	Cu	0.12–0.13
Salts			
Oncorhynchus mykiss	6 h-LC_{40}	NaCl	20,000
Daphnia magna	24-48 h-LC_{50}	NaCl	1020–3240
Daphnia magna	48 h-LC_{50}	Cl_2	0.1–0.2
Daphnia pulex	48 h-LC_{50}	Cl^-	2042
%* = percentage concentration.			

Nutrients, salts, metals, antibiotics, among others separately can generate effect at different levels. On the one hand, NH_4^+ within swine slurries reaches values above 1 g/L. Thus, ammonium on cladoceran aquatic organisms (Ceriodaphnia dubia, Moina macrocopa, Daphnia carinata, Monia australiensis) generate mortality at values above 7.5 mg NH_4^+/L. Indeed, ammonium causes enzymatic inhibition and cell disruption. Moreover, NH_4^+ is instable transforming to NH_3 under alkaline pH (>8) and/or temperature increasing. Indeed, Daphnia magna shows acute toxicity at ammonia concentrations above 2.9 mg/L. Meanwhile, organisms of higher trophic levels (Penaeus semisulcatus) reports acute toxicity at concentrations from 11 to 55 mgNH_4^+/L. In the soil, toxic effects of NH_4^+/NH_3 have been observed mainly in vegetable species, generating necrosis, reduction/stimulation of growth and sensitivity to frost. Indeed, Lactuca sativa and Hordeum vulgare report chronic effects (growth decreasing) at concentrations from 0.002 mg NH_3/L.

Table: Ecotoxicological characteristics terrestrial organisms.

Organism	Test	Compound	Concentration (mg/L o % *)
Swine purines			
Lepidium sativum L.	Growth 24, 48, 72 h	Raw swine slurry	3–10*
Eisenia foetida	Growth 28d	Untreated purine	25*
Nutrients			
Lactuca sativa	Growth 7d	NH_3	0.002
Hordeum vulgare	Growth 8d	NH_4^+	0.18
Metals			
Eisenia foetida	48 h-LC_{50}	Zn	268–439

Eisenia foetida	48 h-LC_{50}	Cu	153–249
Salts			
Different kind of plants	Growth	NaCl	0–1280
%* = percentage concentration.			

Chlorides can exceed 1 g/L within swine slurries, being reported less acute toxicity of NaCl (1020–3240 mg/L) on Daphnia magna than Cl_2 (0.1–0.2 mg/L). However, toxic effect is decreased at high trophic levels (the chlorinated compounds toxicity decreases on higher trophic levels (Oncorhynchus mykiss)), reaching concentrations above 20 g/L. In terrestrial environments, chlorides can cause chronic effects on the vegetal germination. Swine slurries are compound also by micronutrients (copper and zinc), which not exceeds 0.5 g/L. However, acute toxicity on aquatic organisms (Danio rerio) reaches concentrations about 0.1 mg/L and on terrestrial (Eisenia foetida) it is not exceeding at values of 249 mg/L.

Olfactometry Characteristics

The odoriferous characteristics within swine slurries are evaluated using analytical methods (compounds) and sensory methods (odor). Olfactometric characteristics measured as odor concentration (OC/m^3) are influenced by the type/time of slurries storage. Indeed, fattening phase (2.1–5.6 times) has more odor than maternity/weaning. On the other hand, fresh swine slurries (1.5–5.4 times) generate more odor than stored slurries (2 months). These results have also shown the odor relationship with volatile compounds presence, which from by diet or slurries management. Thereby, sulfur, ammonia, phenolic and volatile fatty acid compounds from swine slurries influence the odor generation, which show different correlation degree (R^2 = 0.66–0.89) with the generated odor. Particularly, odor precursor compounds from the swine production will be given by the protein's presence from food, generating sulfurous, indole, phenolic and long-chain fatty acids.

Meanwhile, carbohydrates generate short-chain or volatile fatty acids (less than 6 carbons). Thus, biological and chemical conditions determine the odor compounds formation. It has been possible to establish the presence of some autochthonous microorganisms (Streptococcus, Peptostreptococcus, Eubacterium, Lactobacilli, Escherichia, Clostridium, Propionibacterium, Bacteroides, Megasphaera, among others). Incomplete anaerobic digestion from these microorganisms produce and/or reduce macromolecules to odor compounds, so intermediates (e.g. volatile fatty acids) and finals (e.g. sulfides). However, not only microorganisms can condition the odorant compounds formation, but also environmental conditions (pH and temperature), which active biological and chemical (e.g. ammonium/ammonia) processes. Temperature affects the microbial growth rate; while, pH influences the buffer capacity, favoring volatile fatty acids generation.

Table: Olfactometric characteristics from swine slurries.

Type of waste	Olfactometric technique	Odor threshold (D-T)*, Odor concentration (OC/m^3)^ or Odor index (OI°)
Bovine manure	Field (Nasal Ranger, box)	8.6–157.7*
	Dynamic	124.2–6561^
Sheep, bovine and pig manure	Field (Nasal Ranger)	<2–60*
Swine purines	Dynamic	120–792^
Swine manure	Triangular odor bags (Dynamic mod.)	26.2–58.7°
Porcine, avian and bovine manure	Dynamic	<1–1000^
*=D-T units, ^ = OC/m^3 units, ° = OI units.		

Reduction and Re-valorization of Swine Slurries

Slurries Reduction

Swine has low digestibility (<30%) of nutrients and micronutrients, being necessary mechanisms of digestibility improvement, which could improve the physicochemical characteristics from slurries (feces + urine). Indeed, about 78% of N and P from swine food (proteins) is not assimilated, excreting concentrated urine and feces. Therefore, farm managements are focused on the improvement of diet type and food quantity during each phase growth. On the one hand, raw protein is substituted by fiber, reducing until 8% (10 g RP/kg food) of nitrogen in the urine. Other strategies are related to vary crude protein concentration (155, 145 and 135 g RP/kg) in the food, achieving the decrease of NH_4^+ (20.3–28.4%) in the excreta. Studies have evaluated the replacement of crude protein by digestible or ileal amino acids (lysine, threonine, methionine, tryptophan, isoleucine and valine), finding that they can reduce the ammonium excretion in the urine from 40 to 50%.

Meanwhile, other techniques use feeding multi-phases, which improve the protein digestibility, reducing between 20 and 42% the nitrogen excretion. On the other hand, introducing phytases in the swine diet, it is possible to reduce 18% of phosphorus in the feces. Metals (Cu, Zn) used as growth promoters have been decreased (100–250 ppm Cu, 2000–3000 ppm Zn) by antibiotics (3–220 g/ton food). Indeed, sulfonamides, tetracyclines and β-lactams increase the index conversion rate between 3 and 4%, improving the protein assimilation. However, antibiotics also are excreted up to 10%, not being a good strategy because they are emerging contaminants. The implementation of efficient water drinking reduces the slurry generation. The dozers incorporation and excreta handling techniques (e.g. hot beds) could reduce the floor washing, reusing waste organics (rice husk, straw). These strategies have reached reduce water requirements from 5 to 80%.

Treatment and Re-evaluation of By-products

The swine slurry treatment is the most used tool management within intensive farms by environmental pressures (legislation), which regulates its discharge on water bodies or soil revaluation. The slurries management requires a balance between the environmental/social and economic requirements in the farms. Ideally, this management starts with the excreta fractionation (slurry = urine + feces/feces) by physical/chemical separation. Some techniques, such as: polymers, filter press, flotation, sedimentation, screw press, among others are used to remove sedimentable/suspended material, reducing mainly organic matter (62–84%) and phosphorus (70–89%). The solid fraction corresponding to non-mineralized organic matter is subjected to composting (aerobic/anaerobic), which stabilizes giving it agronomic properties (C/N < 20). The liquid fraction (slurries) with a C/N ratio about 10 is subjected to biological (aerobic/anaerobic) removal processes of organic matter, nutrients and other micro contaminants (metals, emergent).

Several technologies are grouped within the aerobic biological processes (aerated lagoons, activated sludge, among others) and anaerobic (anaerobic lagoons, fixed bed reactors, SBR or Sequencing Batch Reactor, UASB or Upflow Anaerobic Sludge Blanket, among others). Thus, the removal of dissolved and colloidal organic matter is usually carried out by anaerobic lagoons (Environmentally Superior Technologies), which reduce 50% of organic matter. Meanwhile, anaerobic reactors (Manure-based biogas plants) remove more than 80% of organic matter. The by-products obtained from this stage are usually stabilized effluents (C/N < 10) used as soil stabilizer and biogas. In this last point, specific temperature conditions (psychrophilic, mesophilic and thermophilic) have allowed the biogas (60–70% CH_4, 40–30% CO_2) production between 0.03 (anaerobic lagoons) and 650 (anaerobic reactors) m^3/d. The main biogas uses are related with thermal energy (0.02–390 m^3 gas) and/or electrical (0.07–1560 kWh) within farms. Additionally, anaerobic treatment under optimum conditions (35 °C) reduces odors (1.9 units depending on the hedonic tone). This anaerobically treated effluent can be subjected to biological treatment (nitrification/denitrification, SBR or constructed wetlands) or physical–chemical (stripping, vacuum evaporation, precipitation) to nutrients removal (nitrogen and phosphorus). The nitrogen removal efficiencies vary from 40% (constructed wetlands) to 97% (denitrification–denitrification) and 100% (stripping). Meanwhile, phosphorus is removed between 44% (constructed wetlands) and 80% (chemical precipitation). In very few cases, have been reported metals removal (Cu, Zn), mainly due to their low concentrations (<1 g/L). However, the metals removal has allowed obtaining removal efficiencies between 75% (precipitation) and 92% (constructed wetlands). The by-products obtained in this stage can vary from crystallized ammonium salts for agronomic use to treated effluents with a C/N ratio <5 usable in irrigation.

The irrigation (slurries) or soil stabilization (solid) are the most used re-valuation techniques. The treated swine slurries have nutritional value (N:P:K: 1:0.6:0.4–1:0.3:1) to be used in cereals irrigation for swine consumption (1.2:0.2:1 a 1.6:0.3:1). Under optimal irrigation conditions (150–200 kg N/ha year) some soil characteristics with agronomic importance (organic matter content and moisture retention) could be improved. The slurries re-valorization in irrigation has decreased the chemical fertilizers use, being in some countries (New Zealand) valued economically (21 million USD/year). However, the livestock production intensification vs. land availability (Europe) has carried out to optimize the nutrients recovery. Thus, it is necessary to consider within slurries the balance of macro (N:P: K) and micronutrients (metals), as well as other contaminants (pathogens, emergent). Moreover, this balance must consider soil nutritional requirements and crop type, according to the international revaluation legislation. These tools would allow the cadasters establishment, which could facilitate the communication between nutritional requirements sites (agricultural soils) and nutrients-generating sites (animal farms). Preliminary Chilean studies have reported opportunities of livestock slurries revaluation in agriculture relating nutrients recovery sites with adequate agriculture sites.

The biogas production from anaerobic digestion of swine slurries is another revaluation alternative. In Europe, technologies based on biogas generation have been favored by state subsidies (10–30% investment cost) at farms level. These initiatives have allowed the building the more than 5000 anaerobic digestion projects. Swine slurries have potential capacity of methane generation between 0.25 and 0.30 m^3CH_4/kg VS. Meanwhile, co-digestion with lignocellulosic materials (e.g. crop waste) could increase more than 16% produced methane. Thus, centralized plants of anaerobic co-digestion are the current trend within European agricultural sector. Successful experiences have been reported in Germany and Denmark, where more than 32 plants have been built with a capacity between 16,000 and 200,000 tons/year of waste and with a production between 0.7 and 5.7 million m^3 biogas/year. Preliminary Chilean studies have allowed establishing the potential generation of bioenergy (biogas) from anaerobic co-digestion (livestock/crop wastes) at regional level. Figure describes excretes management in the swine production.

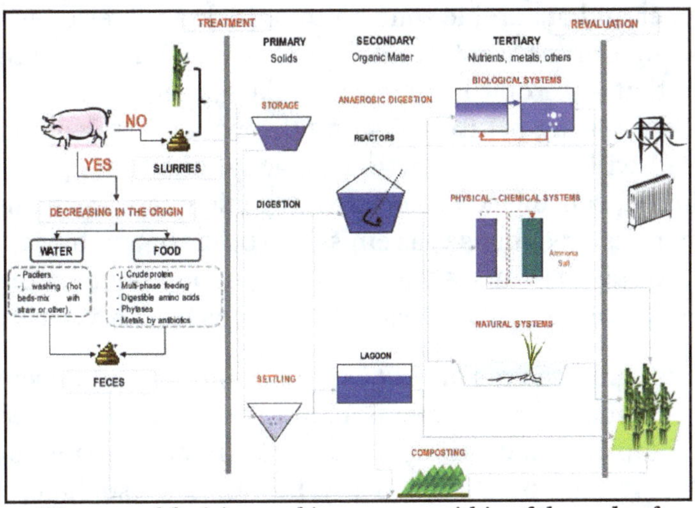

Diagram of decision-making process within of the cycle of generation, treatment, and revaluation of swine slurry.

Swine Slurries Treatment Technologies

Organic Matter Removal

Anaerobic treatment technologies are widely used in this field, because they allow removing organic matter, pathogens, odors, generating by-products (biogas, bio-fertilizer). There several anaerobic technologies, such as: biodigesters, anaerobic lagoon, complete mixing anaerobic reactors, anaerobic filters, UASB or Upflow Anaerobic Sludge Blanket, among other. Worldwide, anaerobic technology based on conventional anaerobic mesophilic reactors (4935 plants) and thermophilic reactors (321 plants) treat around 49 million ton/year slurries, which have been built in Europe, mainly in Germany (more than 70%). State subsides, innovation technologic and centralization of biogas plants using agricultural wastes (livestock, crop) have favored their implementation. Meanwhile, more than 7000 technologies based on anaerobic lagoon have been built, but with only 60% operationally actives in the United States. The conventional anaerobic reactors (complete mixture, piston flow) have been limited by its costs, remaining in disuse or only in building project around 77%. However, currently these technologies are being replaced by more efficient technologies in terms of removal of nutrients, odors and pathogens. Technologies such as tubular biodigesters are the most used in countries where the livestock production in non-intensive.

The anaerobic lagoons are extensive and conventional typology, where organic matter is biodegraded without hydraulic or thermal control (environmental conditions). In the livestock sector, these systems offer some advantages related to their storage capacity and operation easy. However, they can also cause odors, requiring spaces far spaces from neighboring population. Anaerobic lagoons obtain organic matter removal efficiencies between 26 and 79%, working under environmental conditions (5–32 °C) and with residence times from 90 to 232 days. The lagoon design must consider thermal vertical stratification, suggesting depths between 0.8 and 4.0 m to maintain facultative conditions upper and anaerobic zone bottom. Its longer residence time also favors the macro (organic matter, nutrients) and micro (metals) nutrients precipitation. Thus, has been observed that anaerobic lagoon accumulates more than 50% COD, TN, P and Cu in the bottom, decreasing pH from 6.5–7.2 throughout its depth (>1.5 m). Another advantage is related with their disinfection capacity, due to its prolonged time exposure to solar radiation, causing cell lysis of pathogenic microorganisms. However, this time exposure generates greenhouse gas emission (0.02–0.5 m^3 biogas/m^2 d) and odors (168–262 OC/m^3, 101 µg NH^3/m^2s, 5.7 µg H_2S/m^2 s). Operational improvements mainly related to the emission of gases from anaerobic lagoons have been made covering them.

The conventional full mix reactors or CSRT (Continuous Stirred Reactor Tank) are controlled systems, where the hydraulic retention time (residence time) is equal to the cell retention time. The complete mixture is achieved through the recirculation from 25 to 40% of biogas generated. It can operate under psychrophilic, mesophilic, and thermophilic conditions, which are carried out in two stages (acidogenic and methanogenic reactors).

This technology has been widely used mainly in Europe, obtaining organic matter removal efficiencies from 25 to 74%, but with organic loading between 5 and 40 times greater than anaerobic lagoon. CSRT reactor has been improved with the recirculation.

Other conventional technologies as anaerobic filters or AF generate biofilm around of the material support surface, while the flow goes up throughout the filter. Inert (nylon meshes, polyurethane foams, polypropylene rings) and organic (blocks, wood chips) support material have been used. The main advantage of this technology is that can operate at organic loading between 69 and 142 times greater than anaerobic lagoons; but have clogging problems.

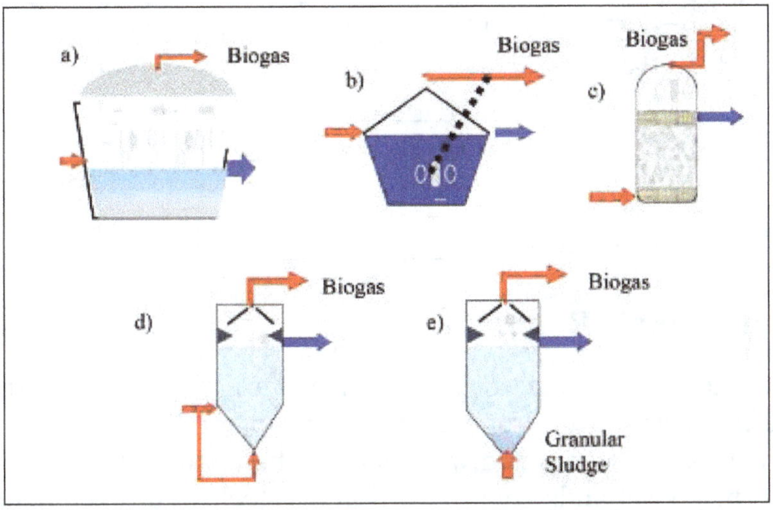

Schemes of different types of anaerobic reactors used in the swine slurry treatment: (a) Lagoon, (b) CSRT, (c) AF, (d) AFBR, (e) UASB.

The most advance technology has been developed to improve operational problems of conventional technologies. On the one hand, AFBR systems (Anaerobic Fluidized Bed Reactor) are technologies studied mainly at laboratory scale. They use support material (clay, wood and PVC), which is suspended due to the recirculation of the flow. These characteristics partially avoid clogging. On the other hand, UASB technology has been applied at the laboratory and pilot level. These treatment units generate biomass granulated, which sediments (4 m/h) improving the cellular retention time. The flow goes up, favoring the washing of biomass non-granulated the granules are dense, harboring multi-species and diverse microbial communities. However, the granulation processes require longer periods of formation (2–8 months). Other innovative technologies from UASB systems are the EGSB (Expanded Granular Sludge Bed). This last technology is hydraulically improved respect to UASB, because it operates at greater flow velocity (>4 m/h) than UASB. In general, UASB reactors are a viable alternative, since they are considered high load systems, operating at organic loadings between 2 and 162 times higher than conventional systems (anaerobic lagoons, CSRT). The organic matter removal efficiencies reach ranges between 19 and 86%. In addition, they offer other operational advantages related to their volume (0.006–0.5 times less volume than lagoons and CSRT) and sludge production (granular from UASB vs. suspended from lagoons/CSRT). Moreover, UASB generates higher biogas production (0.28–4.05 m^3/m^3 d) than conventional systems (0.02–1.69 m^3/m^3 d), thanks to the fact that they operate at higher organic loading (1–8.1 kg COD/m^3d).

Table: Operational characteristics of anaerobic reactors within swine sector.

Technology	Loading kg (BOD_5*, COD^, VS°) /m^3 d	Temperature °C	Efficiency % (BOD_5*, COD^, VS°)	Biogas m^3 biogas/ kg (BOD_5*, COD^, VS°)	% Methane
Lagoon	0.05–0.08°	24–32	26–79*	0.43–0.80°	86–95
CSRT	0.41–2.04°	20–60	25–74*	0.19–0.83°	60–79
AF	3.44–11.34°	31–55	35–61^	0.03–0.29°	61–87
AFBR	1.1–6.6^	–	66–91^	0.17–0.53^	75–84
UASB	1.0–8.1^	20.7–35	19–86*	0.28–0.50^	54–87

CSRT, continuous stirred reactor tank, AF, anaerobic filter, AFBR, anaerobic fluidized bed reactor, UASB, upflow anaerobic sludge blanket.
* =BOD_5 units, ^ = COD units, ° = VS units.

Nutrients and Metals Removal

Constructed wetlands are used as a cost-effective alternative for the nutrients removal within the livestock sector. In Europe, there are around 60 livestock farms, which treat 17,000 tons/year using constructed wetlands. In the United States, about 33% (~70 farms) of constructed wetlands are used within livestock sector, being mainly (~83%) surface flow constructed wetlands (SF-CW).

Operationally, there is experience in the use of different types of constructed wetlands within swine sector. However, SF-CW is the most used technology, mainly to avoid clogging. Generally, SF-CW are used after the anaerobic lagoon, operating at nutrient loading between 5 and 36 kg N/ha and between 1 and 6 kg P/ha. The nitrogen and phosphorous removal efficiencies obtained reach values from 50 and 90 to and 25–66%, respectively. Moreover, horizontal subsurface flow constructed wetlands HSS-CW have been studied at laboratory and pilot scales, operating at nutrient loading between 69 and 252 kg N/ha d and between 15 and 47 kg P/ha d. The nitrogen and phosphorus removal efficiencies vary from 10 to 24 and 47 to 59%, respectively. There are also experiences hybrid systems (SF/HSS/VSS) operating with plant species emergent and floating, which have achieved nitrogen and phosphorus removal efficiencies higher than 50 and 42%, respectively. In general, constructed wetland systems will be operationally work with any previous technology. Thus, activated sludge has been used prior to construct wetland increasing from 2 to 20 times the nutrients removal than anaerobic lagoons. The metals removal efficiencies have been reported in SF-CW, which have operated at rates from 0.09 to 0.25 kg Cu/ha and 0.58 to 1.58 kg Zn/ha d, obtaining removal efficiencies of up to 83 and 92%, respectively.

Table: Operational characteristics of constructed wetlands within swine sector.

Species	Type	Prior Treatment	HRT (days)	Loading rate		Efficiency (%)	
				kg N/ ha d	kg P/ ha d	N	P
Tl, Sa	6 (HSS-lagoon-HS)	Anaerobic Lagoon	18	7–40	3.0–22.0	37–51^	13–31^
Sa, Sc, Sv, Je, Spa, Ta, Tl	4HSS	Anaerobic Lagoon Nitrifying unit	11–13	4.8–27.2 44–51	0.9–6.0 7–9	50–84^ 78–88^	25–38^ <10^
Ec	1HSS	Activated Sludge	27.5	69.0–262.0	15.0-47.0	10.0-24.0^	47.0-59.0^
Pc, Pa, Tl	Hybrid (2VS + 1HSS)	Anaerobic lagoon + Sand filter Recirculated treated purine (25–100%)	– –	214 –	30 –	50^ 54–67^	42^ 47-49^
Pa, Tl, Cp, Ca, Sl, Sm, Fu, Is, Ma, Ac, Se, Apa, Ls, Rl, S, P, Lm, Sta, Msp, Eca	Hybrid 8(HS+ HSS + VS)	Activated sludge	–	1.2–10	0.4–1	99.3^	99.6^
Gm, Ga, Gs, Mm, Poa	16HSS	Anaerobic digester	–	11.2– 36.0	1.1– 1.8	78–90*	56–66^
Cdp, Ap, Tl	2HSS	Filtration Tank	4.8	93.3	22.1	–	–
Ci, So, Pa	Híbrido (3 VS + 1 HSS)	Aeration lagoon	4–5	76.3	2.6	64^	61^

Currently, there are some innovations related to the constructed wetland treatment. Likewise, the partial recirculation of pre-nitrified slurry has allowed increasing the nitrogen removal via denitrification up to 4 times, decreasing the ammonium volatilization. Other design concepts are based on the use of “marsh-pond-marsh”. Constructed wetland technologies, could be improved operationally using intermittent hydraulic rate, which favors the oxygenation improving the nitrification.

Rabbit Farming

Cuniculture is the agricultural practice of breeding and raising domestic rabbits as livestock for meat, fur or wool. Cuniculture is also used in the form of rabbit lovers and fans to the development and improvement of the breed of rabbits and exhibitions of these efforts. Scientist's cuniculture practices in the use and management of rabbits as a medical model in scientific research. Cuniculture is practiced throughout the world since at least the 5th century.

The abundance of ancient rabbits may have played a role in the naming of Spain. Phoenician sailors visiting the coast around the 12th century BC took the European rabbit to the friends of the Cape hyraxes Procavia capensis his homeland. They called his discovery the I-shepan-ham, which means "land of hyraxes". There is a theory, although it is somewhat debatable that corruption is the name used by the Romans, became Spain, the Latin name of the Iberian Peninsula. The domestication of the European rabbit slowly got up from the combination of game-keeping and farming. Among the numerous foodstuffs imported by sea to Rome during its rule in the Mediterranean was the supply of rabbits from Spain. The Romans also imported ferrets for hunting rabbits, and the Romans then gave the rabbits and the habit of keeping a rabbit in Italy, France, and then throughout the Roman Empire, including the British Isles. Rabbits are kept in fenced areas, and in more detail in the game-saves. In the British Isles, these reserves were known as anthills or garths, and rabbits were known as the rabbits, to distinguish them from the same rabbits are a separate species. The term Warren was also used as the name for a place where hares, partridges and pheasants kept under the watch Keeper called a warrener. In order to limit and protect the rabbits, wall or thick hedge can be built around Warren and Warren could be created on the island. Warrener in exercising control over poachers and other predators and to collect rabbits in snares, nets, dogs, such as greyhounds, or hunting with ferrets. With the growth of falconry, hawks and falcons were also used to collect rabbits and hares.

While in the system of the Warren, rabbits were organised, they were not domesticated. The practice of domestication of the rabbit also came from Rome. Christian monasteries throughout Europe and the Middle East kept rabbits, at least from the 5th century. And rabbits can be allowed to wander freely within the monastery walls, was a more common method of employment rabbit courts or rabbit pits. Rabbit court was a walled area lined with brick and cement, while Pete was much the same, just less lined, and hollowed. Individual boxes or burrow space could build a wall. The rabbits will be in the band in these pits or court, and individuals collected when desired for eating or skinning. Of these holes, which do not allow for easy cleaning, power rabbits for breeding, rabbit keepers are moved to separate cages or pens, which were originally made of wood but now often made of metal in order to improve sanitation.

Rabbits are usually kept in the composition of household livestock farmers and people across Europe. Breeding of rabbits including the collection of weeds and grasses for feed, as a rule, those children of home or private house. These rabbits were mainly shared or meat rabbits, not a specific breed, although regional strains and species occur. Some of these strains remain as regional breeds such as Gotland in Sweden, while others, such as the Land Kaninchen, a spotted rabbit of Germany, have become extinct. Another type of rabbit that is standardized in a breed was the Brabancon, a meat rabbit of the region of Limburg and what now Belgium is. Rabbits of this breed were brought to the market in the port of Ostend, intended for the London markets.

The development of refrigerated transport vessels led to the final collapse of the European meat rabbit trade, as the overpopulation of wild rabbits in Australia, can now be collected and sold. In the Brabancon is now considered extinct, although a descendant, the Dutch breed, remains a popular small rabbit for the pet trade. Besides being harvested for meat, properly cooked rabbit is also the economic factor. Both wild rabbits and home rabbit pelts were valued, and it follows that pelts of particular rabbits will be more appreciated. In 1631, price differences were noted between normal skins rabbit skins rabbit is rich in the champagne region in France. This regional type was recognized Darzhan champagne, silver champagne rabbit. Among the first commercial breeds of Angora, which, as they say, can develop in the Carpathians? They made their way to England, where during the reign of King Henry VIII, the laws prohibited the export of long-haired rabbits as a national treasure. In 1723, long-haired rabbits were brought to southern France from the English sailors, who spoke about the animal initially comes from the Angora region of Turkey. Thus there were two different strains, one in France and one in England.

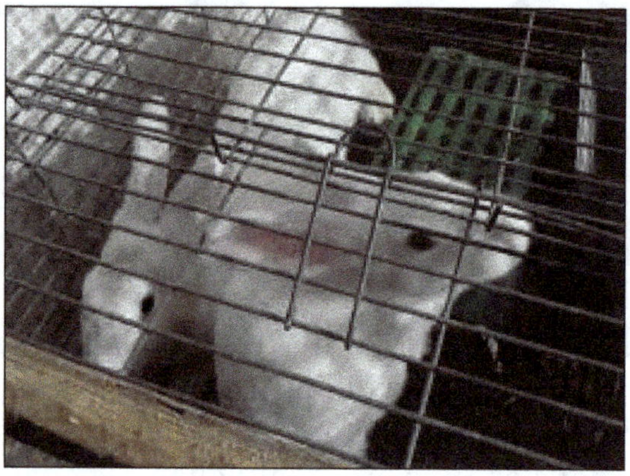

European explorers and sailors rabbits with them to new ports around the world, and brought new varieties to Europe and England with them. With the second voyage of Christopher Columbus in 1494, European livestock was introduced to the new world. Rabbits, along with goats and other hardy livestock often released on Islands for food production for future ships. Introduced occasionally met with disastrous results, as in the devastation in Australia. While cattle and horses were used in the socio-economic spectrum, and were especially prevalent among the rich, the rabbit was saved low-income classes and peasants. This is reflected in the name of the breed that eventually arose in the colonized areas. From Duromo Santa Mountains of Brazil comes the Rustico, which is known in the USA as the Brazilian rabbit. For Creole Rabbit comes from Mexico.

With the growth of scientific animal breeding in the end of 1700 years, is led by Robert Bakewell in particular, different breeds of cattle were developed for specific purposes. Rabbits were one of the last Pets to have these principles applied to them, but rabbits are fast reproductive cycle marked the progress of selection in a short period of time. In addition, rabbits can be kept in a small area, with one person taking care of 300 breeding on an acre of land. Rabbit breeds were developed by individuals, cooperatives and national breeding centers. To meet various production goals, rabbits were exported all over the world. One of the most notable imports of the event was the introduction of a Belgium breed of rabbit from Europe to the USA, beginning in 1888. This led to a brief boom in rabbit breeding, selling and speculation, when the quality of a breeding animal could bring $75 to $200. For comparison, the average daily wage at the time was about 1.00$. In 1900, animal-exports the company recorded 6.000 rabbits successfully supplied to USA and Canada.

Science plays a different role in the upbringing of the rabbit, this time with the rabbits themselves as instruments used for scientific progress. Since Louis Pasteurs experiments in rabies in the second half of the nineteenth century, rabbits have been used as models to study various biomedical problems, including the transmission of disease and protective antiserums. The production of quality animals for meat sale and scientific experimentation has led a number of advancements in rabbit husbandry and nutrition. While the early breeders were limited to local and seasonal foods, which does not allow the maximization of production, health, growth, by 1930 the researchers conducted experiments in the feeding of rabbits, as in experiments, that was isolated vitamins and other nutritional components. This eventually led to the development of different recipes for pelleted rabbit diets. The gradual refinement of the diet led to the wide availability of pelleted feeds, which increase productivity, reduce waste, and improve the health of rabbits, in particular, the health of the mother and breeding.

The final stage of breeding rabbits for meat, wool and fur, and laboratory use - was the breeding of unusual animals as Pets and curiosity. The term of was originally applied to the LOP eared rabbits, as they were the first type to be displayed at the exhibition. The rabbits were first admitted to agricultural shows in England in the 1820-ies, and in 1840 the club was formed to promote and regulate the exhibitions of the "fancy rabbits". In 1918 a new group created with the aim of promoting fur breeds, originally just Beveren and Havana rocks. The club eventually expanded to become a British Council rabbit. Meanwhile, in the United States, clubs that promote different breeds was founded in 1880, and the National securities Association of pet were formed in 1910. This organization will be the American Association of rabbit production. Many thousands of rabbit shows take place each year and are sanctioned in Canada, Mexico, Malaysia, Indonesia and the United States by the Arba.

With the advent of national-level organizations, breeders, the basis for the creation of breeds and varieties using generally accepted standards, keeping and breeding rabbits at exhibitions began to rapidly expand. Such organizations and associations were also created throughout Europe, especially in Germany, France and Scandinavia - allowing for the recognition of local breeds, many of which have similar properties across national borders, and to preserve supply during such interruptions in the First World War and the Second World War. Closely with exhibitions for breeding and fur was raising rabbits for the pet trade. And rabbits kept as companions for centuries, selling rabbits as Pets began to grow in the second half of the twentieth century. Perhaps it was partially because rabbits require less physical space than dogs or cats, but without the specialized habitat of the goldfish. Several breeds of rabbits such as the Holland Lop,

Polish, Dutch dwarf and the critics - were specially bred for the pet trade. The features common to many popular pet breeds of small size, "dwarf" or neotenic features, plush or furry coats, and a variety of colors and patterns.

Outside of the exhibition circles, rabbit raising remained a small but persistent domestic and economic initiatives, in many places not regulated by the rules that regulate the production of cattle. With the continuing urbanization of the world population, rabbit raising gradually reduced, but saw springs in Europe and North America during the Second world war, in conjunction with victory gardens. Eventually, farmers across Europe and in the United States began to approach cuniculture with the same scientific principles as already applied in the production of grain, poultry and hoofstock. National agriculture, breeding station was created to improve the local rabbit strains and to introduce more productive breeds. National breeding centers focused on developing strains for production purposes, including meat, skins and wool. These gradually faded from prominence in the United States, but remain more viable in Europe. Meanwhile, raising rabbit in the local market has gained prominence in developing countries as it is an economical way to obtain protein.

Different agencies for rendering assistance to encourage the use of rabbits as livestock. Animals are particularly useful in areas where women have limited employment outside of the family because rabbits can successfully be stored in small spaces. These same factors contributed to the increased popularity of rabbits as yard of cattle from locavores and settlers in the more developed countries of North America and Europe. Add rabbits to the list of species under threat of extinction heritage which is stored at the protection of livestock has also led to increased interest of ecologists cattle. In contrast, in Asia and particularly in China, Rabbits are more questions and sold for export around the world. World rabbit science Association, founded in 1976, was created to "promote in all possible ways the exchange of knowledge and experience between people in all parts of the world which contribute to the development of various industries in the rabbit." In the WRSA will organize a world conference every four years.

Nearly 1.2 billion rabbits are slaughtered for meat worldwide. In recent years, in some countries, cuniculture has come under pressure from animal rights advocates on several fronts. The use of animals, including rabbits, in scientific experiments has become the subject of increased attention in developed countries. Tighter regulation has increased the cost of production of animals for this purpose, and other experimental options more attractive. Other researchers have abandoned investigations which required animal models. Meanwhile, various rescue groups under the house Rabbit society umbrella have taken an increasingly tough stance against any breeding rabbits even as food in developing countries on the grounds that it contributes to a number of mistreated, unwanted or abandoned animals.

The growth of homesteaders and small holders has led to an increase in the visibility of the rabbit raisers in the geographic areas where they have not attended previously.

This has led to zoning conflicts over the rules of carcasses and disposal of waste. Conflicts also arise with the house rabbit society, as well as ethical vegetarians and vegans regarding the use of rabbits as meat and fur animals, not as Pets. Conversely, many settlers cite concern for the welfare of animals in intensive farming of beef, pork and poultry as a significant factor in the selection was to raise rabbits for meat. Specific future direction cuniculture is unclear, but seems not to be endangered in any particular part of the world. Variety of applications as well as universal type, it is sufficient to keep rabbit rearing foreseeable future, in one aspect or another around the planet.

Aspects of Rabbit Production

Meat Rabbits

Rabbits were raised for meat production in different conditions around the world. Small operations in the yard are still rife in many countries, while large-scale commercial operations are concentrated in Europe and Asia. For small businesses, many local breeds of rabbits may be easier to use. Many local, Rustico, Landrace or other heritage breeds can be used only in a certain geographical area. Unsatisfactory or culling animals from breeding other purposes can also be used for meat, particularly in small operations. Paradoxically, giant rabbit breed is rarely used for meat production due to an elongated growth that lead to high costs of feed and their large size bone tissue, which reduces dress in percentages. A dwarf breed is also rarely used, due to high production costs, slow growth and low seed rate. Unlike many breeds and types that are used in small operations, breeds such as New Zealand and California, along with hybrids of these breeds commonly used for meat in commercial rabbitries.

The main qualities of a good meat rabbit breeding cattle are growth rate and size at slaughter, as well as a good ability to be a mother. Specific areas for commercial species have been developed and make full use of these quality Bunnies can be killed for seven weeks and makes these strains usually cause litters of 8 to 12 kits. Other breeds of rabbit developed for commercial production of meat include the Florida white and Altex. Rabbit breeding stock rose in France, especially popular with rabbit meat farmers at the international level, some of them bought as far as China to improve the local herd of rabbits. Large-scale operations attempts to increase revenues, balancing land use, labour, animal health, and investment in infrastructure.

Specific infrastructure and strain qualities depend on the geographical area. Operation in urban areas may emphasize odor control and space allocation, stacking cages on top of each other, with an automatic cleaning system that throw feces and urine. In rural subtropical and tropical areas, temperature control becomes an increasingly important issue, and the use of air conditioned buildings is widely distributed in many areas. Schedule rabbit breeding depends on the individual operation. Before the development of modern balanced diets of rabbits, rabbit breeding was limited food available in DOE. Without adequate calories and protein, the DOE would not be fertile, would abort and re-absorb a fetus during pregnancy, or to put a small amount of weak sets. In these conditions the female is re-bred soon after weaning her last litter when the kits are two months old. This allowed for a maximum of four litters per year.

Advances food, such as those published in research station USDA Rabbit, has led to increased health for breeding and survival of young animals. Besides offering an excellent, balanced diet for growing sets allowed for better health and less disease among the animals to slaughter. Current practice includes the possibility of re-breeding the female within a few days of delivery in accordance with the behavior of wild rabbits in the spring /early summer, when forage availability is in full swing. This can lead to eight or more litters per year. NOO the perfect meat-stock genetics can produce five times her body weight in the fryer year. Criticism is more severe schedule selection was made on the grounds that the re-pairing, which is closely overly stressful for the DOE. To determine the consequences for the health of the schedule of reproduction is complicated by domestic rabbits reproductive psychology – unlike some other mammalian species, rabbits are more likely to develop cancer of the uterus when they are not used for breeding, are often bred than when. In efficient production systems, Rabbits can turn 20 percent of the proteins they eat into edible meat, compared to 22 to 23 percent for broiler chickens, 16 to 18 percent for pigs and 8 to 12 percent, on beef, rabbit meat is more economical from the point of view of nutritional energy than beef. "Rabbit fryer" rabbits that are between 70 and 90 days, weighing in at 3-5 pounds with a 1.4–2.3 kg live weight. "Rabbit roasters" rabbits from 90 days to 6 months, weighing 5-8 kg 2.3–3.6 kg of live weight. "Rabbit stewers" rabbits 6 months and older, weighing more than 8 pounds, 3.6 kg "Dark fryer", i.e. any color except white usually garner a lower price than the "white powder", also called "albino fryers", because of the slightly darker hue of meat. Purely pink carcasses are preferred by most consumers. Dark fryers also de harder to hide than the white powder.

In the US, White fryers garner the highest prices per kilogram of live weight. In Europe, however, a significant portion of the market remains in the dark fryers, which come from old and large rabbits. In the kitchen, deep fryers usually prepared differently from the white fryers. In 1990 the world's annual rabbit meat production is estimated at 1.5 million tons. In 2014, the population was estimated to be 2 million tons. China is one of the world's largest producers and consumers of rabbit meat accounts for about 30% of the global consumption volume. In China, rabbits are grown in many provinces, with the majority of rabbit meat is about 70% of the national production, equaling each year, about 420.000 tons consumed in the Sichuan basin the Sichuan and Chongqing, where he is especially popular. Well-known chef mark Bittman wrote that rabbit "tastes like chicken", because both "blank palette on which we can layer any flavor we like.

Wool Rabbit

Rabbits such as the Angora, American fuzzy lop, and t-shirts. However, since the American fuzzy lop and Jersey wooly are both dwarf breeds, only the much larger Angora breeds such as the English Angora, satin Angora, giant Angora and French Angora are used for the production of wool. Their long fur is sheared, combed, or plucked gently pulling the hairs from the body during molting, and then spun into yarn used to make different products. Angora sweaters can be purchased at many clothing stores and is usually mixed with other types of wool. In 2010, 70% Angora rabbit wool has been manufactured in China. Rabbit wool, called Angora, is 5 times warmer than wool sheep.

Rabbits Fur

Several breeds of rabbits were developed from the fur trade in mind. Breeds such as the Rex, Satin, chinchilla are often raised for their fur. Each breed has characteristics of fur and has a wide range of colors and patterns. Fur rabbits are fed a diet specially balanced for the production of fur and pelts are harvested when they have reached perfect condition. Rabbit fur is widely used throughout the world.

Exhibition of Rabbits

Many breeders breed their rabbits for competition amongst purebred rabbits of the same breed. Rabbits are judged according to the standards put forward by the governing organizations of a particular country. These associations, being made up of people, may be distinctly political and reflect the preferences of certain persons in the governing bodies. However, as mechanisms for the preservation of rare breeds of rabbits, to enable communication between breeders and encourage the education of the population, these organizations are invaluable. Examples include the American Association for the rabbit and the British Rabbit Council.

In the Field of Crop Production

Modern methods for housing domestic rabbits vary from region to region around the world and by type of rabbit, technological & financial opportunities and constraints, intended use, number of animals, and the particular preferences of the owner/farmer. Various purposes like to increase the number of animals per unit area, especially in areas with high land prices or a small living area minimize labor, reduce costs, increase life expectancy and the health of the animals and meet the specific needs of the market, for example, for the cleaning of wool, or rabbits raised on pasture. Not all of these objectives are complementary. Where keeping rabbits is regulated by governments, specific requirements were enacted. Different industries also have common practices that produce predictable results for this type of rabbit product.

Extensive Cuniculture Practices

Extensive cuniculture refers to the practice of accounting rabbits at a lower density and a lower level of production than intensive culture. In particular, with regard to rabbits, this type of production was nearly universal before the germ theory understanding of infectious parasites especially coccidia and the role of nutrition in the prevention of abortion and reproductive losses. The biggest rabbit "to keep the" methods will be the harvest of wild or feral rabbits for meat or fur market, such as occurred in Australia before

1990-ies. Warren-based cuniculture somewhat more controlled, as the animals, usually kept in a certain area and a limited number of available foods. Finally, there have been developed various methods of raising rabbits with pasture as the main source of food.

Herding rabbits into a fence, not a cage also known as the colony in the field of plant, not often conducted due to high mortality from weather and predators. More usual but still rare in absolute terms, the rabbits and the practice is the practice of keeping rabbits in a mobile cage with an open or slatted floor so that the rabbits can access grass but still be close at hand and protected from the weather and predators. This method of growing rabbits, as a rule, not as a result of an overall reduction in the need to Supplement feed. The growing period to market weight a lot more for grass fed, not pellet fed animals, and many manufacturers continue to offer a small number of complete rations during the growing period. Cages or enclosures for this type of farming usually made from a combination of wood and metal wire, made portable enough for a person to move the rabbits daily to fresh ground, and of such a size to hold a litter of 6 to 12 baby rabbits at the market weight of 4 to 5 pounds. Protection from the sun and rain are an important health problem, as resistance against the attack of predators and ability to be cleaned in order to prevent losses from coccidious.

Intensive Cuniculture Practices

Intensive cuniculture is more focused, effective and sensitive using a higher density of animals in space and a higher turnover. How much work is required for the production of each harvested hide, a kilogram of cotton or fryer market - and the quality thereof - may be higher or lower than extensive methods. The successful operations to raise healthy rabbits that produce durable goods run the Gamut from thousands of animals to less than a dozen. Simple cells, kitchen floors, and even natural holes can provide shelter from the elements, while the rabbits fed from the garden or given that can be harvested as they grow to produce community food and textiles. Intensive cuniculture can also be practiced in a closed, climate-controlled barn, where rows of cells reliable house rabbits eat pellets and treats to daily health checks or weekly to check the weight. Experts in the field of veterinary medicine and biosecurity may be part of a large-scale operation, while individual rabbits in small installations might be better or worse - care.

Tasks for Successful Production

Some difficulties to account for rabbits vary depending on the specific practice. Losses from coccidiosis are much more common when rabbits are kept on the ground as in warrens or colonies or on solid floors than when on wire or plates cages that keep rabbits elevated away from urine and feces. Grazing rabbits are more susceptible to attack by predator. Rabbits kept indoors with proper temperature rarely suffer from heat loss compared to rabbits placed outdoors in the summer. At the same time, if the rabbits are placed inside without adequate ventilation, respiratory diseases can be a significant cause of morbidity and mortality. The production of feed, are rarely able to raise more than 3 litters in a year without heavy losses from destruction of weak sets of miscarriage, and re-adsorption related to poor nutrition and insufficient protein intake. In contrast, rabbits fed commercial pelleted diets can face losses related to low fiber intake.

Genetics

Studying rabbit genetics is of interest to researchers, enthusiasts, and the fur and the meat industry. Each of these groups has different needs for genetic information. In the biomedical research community and pharmaceutical industry, rabbit's genetics are important for the production of antibodies, testing of the toxicity of consumer products, and models of the study organism. Among rabbit fans in fiber and fur industry, the genetics of coat color and properties of hair are paramount. The meat industry is based on the genetics of disease resistance, feed conversion ratio, and reproduction potential. The rabbit genome was sequenced and is publicly available. Mitochondrial DNA was also sequenced. In 2011, parts of the rabbit genome were sequenced in greater depth in order to identify changes in the genome.

Genetics: Maps Gene Linkage

The first genetic studies focused on the distance relationship between the various gross phenotypes using analysis of adhesion. Between 1924 and 1941, relations with, g, B, du, ru, l, P1, P2, a, DW, W, f and br was created phenotype given below:

- Du -- Dutch flavor.
- P1, P2 genes -- Rex.
- F-the Sphinx.
- Mr. yellow fat
- DG-the gene for dwarfism.
- A-Agouti.
- With albinism.
- -- An English flavor.
- BR -- brachydactyly.
- I. Angora.
- W -- wide average-color stripes.

The distance between these genes is as follows, these chromosomes. The format gene1 - distance - gene2 --

- F -- 28.3 -- BR.

- And -- 14.7 -- DG -- -- 15.4 W.
- P1 -- P2 -- 17.2.
- Do an -- 1.2 -- -- 13.1 -- I.
- With -- 14.4 -- -- 28.4 g-b.

Poultry Farming

Poultry farming is the agricultural practice of breeding and raising birds (typically chickens, turkeys, ducks, and geese) for the purpose of harvesting their meat or eggs for human consumption. Poultry farmers are responsible for the slaughter of tens of billions of birds per year. Also called chicken farms, poultry farms are designed for maximum efficiency. The speed with which these operations "process" chickens often results in extremely painful deaths for the birds.

Layer Hens

In poultry farming, layer hens are the female chickens that spend their entire lives laying eggs for commercial purposes. The operation collects the eggs and packages them for supermarkets and restaurants all over the world.

Broiler Chickens

Broiler chickens are chickens raised for meat. They're part of the poultry farming operation that allows eggs to be fertilized to increase production. However, the chicks that come out of these eggs aren't raised by their mothers but instead incubated before being hatched and moved to mass housing. Chickens, when allowed to live out their natural lives, can reach 15 years of age or more. Broiler chickens are typically slaughtered at 42 days of age. This is because they're bred to mature more quickly than is natural, which means that they often suffer from serious health problems during their short lives.

Various Methods of Poultry Farming

About 74% of the total world poultry meat and 68% of poultry eggs are produced from intensive poultry farming method, from free-range farming methods. Free range farming method is used for a large number of poultry birds with high stocking density. There are some basic differences between the free-range farming and intensive poultry farming methods.

Intensive Poultry Farming

- Intensive poultry farming efficiently saves space, feed, labour and other resources and increases production.
- This system will be fully under the control of the farmer thus ensuring production throughout the year.
- It has a drawback that chances of health risk for the poultry birds are high and regular inspection is needed for healthy birds and quality eggs from them.

Free Range Farming Method

- Free range farming requires more open space; better management and production sometimes get affected by climatic conditions.
- Farmer have to take more care of the flock on unfavourable climatic conditions.
- Requires regular inspection to keep the poultry birds free from any diseases and to produce quality eggs and meat.

Layer Poultry Farming Methods

Majorly egg-laying poultry birds are reared under layer poultry farming method. Commercial hens begin to lay eggs from the age of 12-20 weeks. And they start laying eggs regularly for 25 weeks. The farmer can make a good yield from them at this stage. After 70-72 weeks of age, the egg production stops and the layer poultry production get reduced. Usually, commercial layer poultry farmers keep the hens for 12 months and later use them for meat purposes.

In layer poultry farming systems, the environmental conditions of the farm are often controlled automatically for ensuring maximum production. Light and temperature are the factors which induce the hens to lay eggs. So by increasing the lighting period and maintaining an optimum temperature inside the farm production can be increased. By this commercial layer farming, chicken breeds can produce more than 300 eggs per year. Common layer poultry farming is as follows:

Free Range Roaming

At free range poultry farming, the farmer provides free range roaming facility for the poultry birds for a certain period of the day and they are kept inside the cage at night from predators and adverse weather conditions. Generally, in free-range poultry farming, the chickens roam freely at the day-time which means they spent half of their life outside the cage. This makes them naturally more healthy and resistant to diseases by natural means.

For free-range poultry birds farming, an open space with adequate drainage facilities good ventilation, protection from prevailing winds, predators, excessive heat or cold or dampness. Excessive cold, dampness and heat adversely affect the productivity of the free-range poultry birds. The main advantage of this method is it needs less feed as the chickens roam freely, they find more food naturally. Although free-range poultry farming is very suitable for egg laying poultry birds it has difficulties too.

Organic Method

Organic layer poultry rearing system is also another form of free-range poultry farming. At organic method, a small number of certain poultry birds are reared by organic methods with low stock density. Organic systems have some restrictions in the routine and feeding system. In organic laying systems, the producer should keep the number of poultry birds by 1000 per hector and maximum 2000 birds in each house.

Yarding Method

A method in which cows and hens are raised together is generally termed as the yarding method. Small farmers make a fence in his yard and keep all the poultry birds and cows in it, so they are free to move inside and roam around that space.

Battery Cage Method

It is a very common method popular and used in many countries. In this method, a small metal cage which cans accommodate 3 to 8 hens is used. Generally, the walls of the cage are made of mesh or solid metal and the floor is made sloped to drop the faeces down. When the hen lays eggs, the egg will be collected in egg collecting conveyor belt of the cage. Food is provided in front of the cage by a long bisected metal or semi-circle plastic pipe and water is served to the using overhead nipple system. The cages are arranged in rows and one above another. Usually, there will be several floors kept under one roof which can contain thousands of hens. The light intensity is kept lower than 10 lux to reduce feather and vent pecking. The main benefits of battery cage system are:

- Easy to care birds.
- Easy to collect eggs.
- Requires less feed for maximum egg productivity.
- Thousands of hens can be accommodated in less space.
- Bird suffers less by internal parasites.
- Low labour cost.

Apart from these benefits, the battery cage system has some drawbacks too. By accommodating a large number of birds, the amount of CO_2 accumulation inside the house would be high. The hens don't get sufficient space to walk, flap their wings, stand or perch inside the small cage. By this, they may suffer from frustration and boredom and their behaviours may change which affect the production. Battery cage system is banned in some countries as it is considered as animal cruelty and against animal welfare.

Furnished Cage Method

It can be said as the updated version of the battery cage system. In this system, hens get more space and facilities than the battery cage system. They will get sufficient space to walk, perch, flap their wings, nests, special feed and water pot etc.

Broiler Poultry Farming Methods

Poultry birds which are reared for meat purposes are called broiler poultry. By practicing modern farming methods the chicken becomes consumable within 5-6 weeks of age. Common raising systems which are mostly used for commercial broiler poultry farming are as follows:

- Indoor Raising method: In this method, broilers are kept inside a large and open house (known as grow out houses) with flooring of rice hulls, wood shavings; peanut shells etc. are used as litter. The chicken becomes consumable within their 5-6 weeks of age. These types of houses are well equipped with mechanical systems for food and water delivery to the birds. A well-ventilated system with cooler and heater are must in these houses and it is very important to keep the house clean and dry. A 16000 sq.ft house can accommodate as much as 20,000 birds.

- Free-range Methods: The broilers are kept like the free layer in the free-range method. Broiler breeds which grow slowly (takes more than 8 weeks for reaching slaughter weight) are suitable for this method. The main facilities of free-range farming systems are that it allows the birds scratching, foraging, pecking and outdoor exercise.

- Organic Farming Methods: The organic farming method is almost the same as free-range systems. But the main difference is that in organic farming methods the birds are not allowed for randomly using in-feed or in-water medications, other food additives and synthetic amino acids. This system is very suitable for the poultry breeds which reach slaughter weight slowly (around 12 weeks).

Poultry Farm Equipment

A standard poultry farm uses many tools and equipment. They are meant for housing the birds, care or butchering.

Incubation Equipment

The number of equipment and tools used in poultry farming is huge. Let us start from the hatching of the eggs into small adorable chicks. The equipment list to run a successful hatchery is setter machine to test humidity and temperature. The farmer needs a hatcher or incubator. There are three types of incubators the tunnels, walk in or vertical fan incubators. The turning mechanisms in some incubators require compressed air system.

Hatchery automation equipment is rack washers high pressure pumps and egg transfer machines. Others include hatchery tray washer, chick box washer, waste removal system and ovo vaccination equipment.

Poultry Hatchery Equipment

- Setter machine,
- Tunnel, walk in vertical fan incubators,
- Compressed air system,
- Rack washers,
- High pressure pumps,
- Egg transfer machines,
- Tray washer,
- Chick box washer,
- Waste removal system,
- Ovo vaccination equipment.

Egg Handling Equipment

Eggs are delicate and crack easily if handled wrong. Equipment to handle eggs is hatching egg transfer machines like vacuum egg lifts. The person needs to stock hundreds of egg trays. A light device to check the internal structure of an egg in a hatchery program is the candler.

- Vacuum egg lifts,
- Egg trays,
- Candler.

Brooder Equipment

The small day old chicks that hatch are brooders. They require lots of warmth, light and space. Common tools are reflectors, stoves, gas brooder or electric brooder. More include infrared bulbs, hovers and chick guard. The chick guard is to restrict the movement of the chicks. The farmer could also introduce electric heater with temperature adjust.

- Reflectors,
- Stoves,
- Gas brooder or electric brooder,
- Infrared bulbs,
- Hovers,
- Chick guard.

Feeding Equipment

The feeding equipment is either manual or automatic. The list includes automatic or semi-automatic feeder. You have the adjustable linear feeder, shell grit box, semi-automatic circular feeder.

- Manual feeder,
- Automatic feeder,
- Semi-automatic feeder,
- Adjustable linear feeder,
- Shell grit box,
- Semi-automatic,
- Circular feeder.

Water Equipment

Water equipment is filers, water softeners, water heaters. Pan/jar, plastic basin, automatic water device. Others are manual drinker, nipple drinker. The advantage of manual drinkers is easy establishment of vaccines and nutrients in the water. Vaccination is delivered through a syringe or automatic vaccinator.

- Filers,
- Water softeners,
- Water heaters,
- Pan/jar,
- Plastic basin,
- Automatic water device,
- Manual drinker,
- Nipple drinker,
- Syringe or automatic vaccinator.

Tools

Basic tools are used for cleaning purposes while special ones are made for certain activities. A specialized tool is the beak trimmer, nest box and weighing scale. Cleaning tools are rakes, sprinklers, spray. To slaughter the mature birds some farms have automatic conveyor belt system. They need guillotine, knives and slaughter slab. More are disposable sleeves, hair caps, overalls.

- Beak trimmer,
- Nest box,
- Weighing scale,
- Rakes,
- Sprinklers,
- Spray,
- Automatic conveyor belt system,
- Guillotine,
- Knives,
- Slaughter slab,
- Disposable,
- Sleeves,
- Hair caps,
- Overalls,
- Wheelbarrows.

Housing

The housing depends on the type of poultry farming. Layers are kept in battery cages while broilers stay in deep litter room. Some use coup boxes, barns free range system. The need heating system, ventilation and space.

- Battery cages,
- Deep litter room,
- Coup boxes,
- Barns,
- Free range system,
- Heating system,
- Ventilation system.

Marketing and Transportation

The transport van should have temperature control gauge. This is because chickens are easily stressed and will die. The farmer needs transport boxes to carry the birds.

Poultry Housing and Management

Poultry production has occupied a leading role in the agriculture industry worldwide in recent years. The compound annual growth rate of poultry protein between 2015 and 2025 is estimated to be +2.4%. Asia, South America and Africa characterized by rapid urbanization, poverty and hot climate recorded the highest growth increment in poultry production. The trend of continuous growth of poultry production in those regions is obvious because it remains the fastest route to bridging the protein demand-supply gap.

Extreme weather conditions in the tropical regions of the world have proven generally detrimental to livestock production and is particularly of interest in chicken because of the latter's high sensitivity to temperature change. Just like mammals, the avian species have the ability to regulate their body temperatures by losing or generating heat in response to environmental temperature. If the body temperature of a bird, which normally runs between 39.4 and 40 °C, is allowed to increase, the bird will not perform well. Heat stress in poultry production had resulted in under-nutrition, stunted growth, reduction in egg production and size, laying of premature eggs and even death. This problem is further compounded by the high body heat generated by genetically improved laying birds with increased metabolic activity resulting from the high rate of egg production.

Poultry housing design plays a vital role in the determination of the internal climatic conditions of the house for optimum health, growth and productive performance of the birds. Consequently, the type of poultry housing system employed by the proposed poultry farm is a function of the prevailing climatic conditions of the region where the farm is located. While open poultry house system has been adjudged a good method of housing in the tropical countries because of the simplicity of its construction, ease of heat management and minimal management cost, the controlled housing system is the most common in the temperate regions of the world.

Poultry Birds and their Thermoregulatory Mechanism

Birds are warm-blooded 'homoeothermic' flighty feathered oviparous vertebrates that possess a high metabolic rate, with a normal breathing rate of 40–50 breaths per minute. On the average, birds maintain an internal body temperature of between 39 and 42.2 °C. During hot weather, poultry birds maintain thermo-neutral temperature by losing heat mainly through conduction, convection, radiation and evaporative cooling.

Sensible heat loss through convection, radiation and conduction is only effective if the environmental temperature is below or within the bird's thermoneutral zone. However, evaporative cooling accounts for about 60% of the heat dissipated during body temperature regulation within the thermoneutral zone. Sensible heat loss includes heat loss through opened surfaces such as wattles, shanks and other featherless areas around the neck and wings. Heat loss for body temperature regulation through this process does not alter the bird's behavioral patterns, feed intake, or metabolism. The effectiveness of sensible heat lost is a function of the temperature difference between the bird and its environment.

Where the environmental temperature exceeds 24 °C evaporative cooling (latent heat loss) becomes the major method of dissipating heat in birds regardless the age. Loss of heat through evaporative cooling at temperatures beyond the thermoneutral zone requires the bird to re-direct energy required for growth and development to panting. However, panting can lead to dehydration and respiratory alkalosis because of inadequate water supply and drop in blood pH due to excessive ejection of carbon dioxide. During panting, evaporative cooling occurs when water evaporates from the respiratory system of the bird. However, this can be hindered by high humidity. This is problematic in high humid environments where poultry farmers employ evaporation cooling as the primary method of air-temperature reduction during the hot periods of the year. Increasing the volume and velocity of air moving over birds enhances heat loss in birds due to convection, removal of heat trapped within the poultry house, and reduction of the effect of high humidity on evaporative cooling.

Water is an essential commodity in poultry production for the nutrients it possesses and its impact on feed consumption. Nipple drinkers to provide cleaner water reduce water spillage and labour for drinker cleaning has replaced the conventional open water system. May et al. observed that chicken consumed more water when reared with conventional open water system in an experiment that compared the conventional open water system to nipple drinking. However, when these drinkers were used to rear chicken in a controlled room with air velocity of 0.25 and 2.1 m/s, birds in the higher air velocity with nipple drinkers did not differ from those on open water drinkers, but experienced increased weight gain and better feed conversion than birds at the lower air velocity. Therefore, it is important to provide and maintain the required ventilation to ensure that the poultry house is conducive for the birds to regulate their body temperature by sensible heat loss.

Heat Stress in Chicken

Heat Stress is a general problem in the poultry industry, especially in the production of chicken meat and egg. Heat stress is experienced by chicken when the environmental temperature equals or rises above 26.7 °C. At this temperature and beyond the birds begin to pant and can be detrimental to attaining the bird's optimum growth rate, hatching ability, egg size, egg shell quality and egg production. The problem of heat stress can be further compounded in a hot environment when the humidity rises. Heat stress has been reported to have adverse effect on broilers comfort, growth rate, feed conversion, and live weight gain.

In poultry production, the sudden exposure of birds to high temperature short periods is referred to as acute heat stress while exposure for extended periods is referred to as chronic stress. Chronic stress has deleterious effects on birds reared in open-sided houses, which is commonly used in the tropics. It has been reported to have adverse effect on growth and production efficiency, egg quality, meat quality, embryonic development, reproductive performance, immunity and disease incidence's in broilers, laying hens and breeders.

Effects of Internal Climate Conditions on Chicken

It is important we understand the effect of internal climatic conditions of the poultry house on the birds, how the birds respond to them, and their implications on heat management for poultry production. The information will provide guidance on parameters for the open poultry house architectural design that will alleviate heat stress to ensure optimum poultry production in the tropics. The climatic factors of interest include temperature, relative humidity, air composition and velocity, and lighting condition.

Temperature

There is a huge debate on the ideal temperature range required for the various classes and age groups of chicken to attain optimum production. This could be because of other climatic factors such as humidity and wind velocity, which influence temperature change and previous adaptation of chicken to climatic change. Generally, chicken perform under a wide range of temperature regardless of its class (broiler, pullet or breeder) or age. However, exposure of chicken to high temperature has been reported to hinder the performance in chicken production. It could also be further compounded by increased relative humidity for its negative effect on evaporative cooling.

Ketelaars recommended a temperature of 30–32 °C at chicken height for day old chicks. Thereafter, the temperature should be decreased by 3–4 °C till the chicks are 4 weeks old as shown in Table. Daghir reported that a temperature range of 18–22 °C is required for growing broilers. It was concluded that birds are comfortable when environmental temperature is within the range of 18–24 °C. However, it should be noted that the optimum performance of chicken is dependent on the market value of the product in relation to feeding cost. It is a challenge to maintain the optimum production temperature in the tropics therefore, it is important that the poultry house designer pay considerable attention to temperature change.

Table: Recommended temperature schedule.

Age of chicken (week)	Temperature range (°C)
1	30–32
2	30–26
3	26–23
4	23–20
≥5	20

Relative Humidity

It was reported that internal temperature above 26.7 °C combined with high relative humidity adversely affected the feed efficiency, feathering, pigmentation, and weight gain of chicken. Furthermore, at internal temperature range of 35–37.8 °C the birds' performances were poor regardless of the change in relative humidity. This means higher humidity can improve the performance of the birds at lower temperature. However, humidity must be controlled for it can provide habitat for microorganisms thus, exposing the birds to the threat of disease.

Relative humidity has a strong relationship with temperature change. At the brooding stage, particularly in the earlier weeks the internal relative humidity may be low or too low because of the warming the chicken requires at that age or when the chicks are thirsty or hatched at higher temperature. Soon enough, the internal relative humidity increases because of the water vapor generated by the evaporative cooling act of chicken to regulate their body temperature as they grow. Consequently, ages 3 weeks and above are very critical periods in chicken production regardless the class of chicken. In Oloyo, it was reported that laying birds during brooding and after brooding require a relative humidity range of 60–80 and 50–70% respectively for optimum performance.

Air Composition

The decomposition of bird's fecal material produces unpleasant and polluted gases, which include ammonia, carbon dioxide, methane and hydrogen sulphide. These gases are of particular interest because of their adverse effects on the performance of birds, cages, human poultry houses and the environment at large. Consequently, for optimum production for chicken a concentration level of 25 ppm and not more than 2500 ppm for ammonia and carbon dioxide was recommended. It was recommended for good birds' health management that removal of fecal material from the poultry house should be done frequently to reduce the volume of gas emission.

Air Velocity

High internal temperature can be controlled to an extent by varying the air velocity within the poultry house. Also, Air velocity plays an important role in convectional cooling and the regulation of air quality. In hot climatic regions, it is recommended that the ventilation capacity should be at least "5m³ per chicken per hour, with inlets amounting to 1.5cm² per m³ ventilation". Hulzebosch reported that still air velocity (0.1–0.2 m/s) could be maintained if the temperature remains within 25–30 °C. However, Lacy and Czarick under the same temperature condition reported a better growth rate at 2 and 3 m/s air velocity respectively for broilers.

In the quest to further understand the effect of air velocity on chicken, factored the ages of chicken within the temperature range 25–30 °C with varying air velocity. The study demonstrated that 6 weeks old broilers benefited from increased air velocity of 2 and 3 m/s than 4 weeks old broilers. This could be because of the high temperature required by younger birds at brooding stage.

Lighting

Lighting at early age in birds have little or no effect on hormonal system, it merely aids birds' activeness including feed intake, growth, and physical and physiological activities. Subsequently, increase in lighting periods and light intensity may cause tiredness, cannibalism, immune responses, leg abnormalities and even death.

The lighting program commonly used is the continuous lighting program of 16 hours light and 8 hours darkness and it has proven successful for overall chicken performance. However, it has been reported that alternating short light and dark period known as intermittent lighting enhances chicken performance. The continuous lighting program with a minimum light intensity of 20 lux is recommended at post-hatch stage (1–7 days old) to help the chick adapt to their environment and aid feeding. Consequently, the light intensity is reduced to about 3–5 lux and intermittent lighting system is introduced for easy control of the birds' activeness for better performance and productivity.

Birds reared under yellow, green, and blue light sources have been reported to have improved body weight compared to those reared under red and orange light sources. Lewis and Morris in a review concluded that the birds reared under blue light show docile trait while those reared under red light were more active and aggressive. In addition, it was noted that the red light improved sexual activities in birds.

Poultry Housing System

The importance of the type of poultry housing system employed for chicken production cannot be over emphasized. It protects the birds from the harsh environmental climatic conditions, which may have adverse effect on the chickens' performance and productivity. In a poultry house, the overall heat generated is the sum of heat generated by the birds, the surrounding environment and biodegradation of fecal material. Thus, the type of housing system to be used is a major determinant factor in the type of management to be adopted in the poultry farm.

Naturally Ventilated Open Housing System

The open poultry housing system has been identified with the tropical region for its simplicity, economic implications and ease of management of heat generation within the building through natural ventilation. However, it is prone to the invasion of insect, rodents, birds and other small predators that can disturb the welfare, productivity and performance of chicken. In the quest to alleviate this problem, dwarf sidewalls are raised to the roof eaves with corrugated wire mesh to keep predators away. Also, gutter filled with insecticides to prevent the invasion of insects are built around the house. Discussed below are design considerations to be factored in when designing an open poultry house for optimum poultry performance and productivity.

Building Orientation

In order to reduce the exposure of sidewall to direct to direct sun radiation the poultry house should be orientated in the east-west direction. This is very vital, because heat stress in birds can be hastened when they are exposed to direct solar radiation. Deep litter rearing may allow the birds avoid direct sunlight but this may lead to clustering or overcrowding of birds in an area of the house. Consequently, make cooling difficult and in severe cases this leads to stampede and even death.

House Width, Length and Height

The east-west orientation of a poultry house may reduce the benefit of prevailing winds blowing from east or west. Therefore, Daghir recommended that the width of the building should not exceed 12 m to prevent this problem. In addition, the problem of uneven air exchange rate and temperature within the building is eradicated. Furthermore, the design must factor in the activities and services rendered by poultry farmers and professionals within the building. These activities may include transfer of chicken, feeding, de-pecking, waste management, vaccination, and so on. Therefore, longer pen house could be strenuous to maintain especially when the activities are carried out manually. Doors can be placed at interval of 15–30 m to make for easy circulation and service delivery. Qureshi recommended that for battery cages, it is rather advisable to factor in the number of tiers to be used. Two–tier cage system facilitates easy air exchange within the building whereas three and four tier cage system can be problematic for air exchange. Therefore, it is recommend that rows of cages should not exceed three with center aisles not less than 1.2 m and a minimum height difference of 1 m from the ceiling.

Roof Slope

A roof slope of 45° was recommended because the angle reduces the heat gain of the roof from the direct solar radiation; maximizes the distance of the bird from the heat accumulated under the roof; quick escape of the heat accumulated under the roof through ridge opening, maximizes air space to improve air exchange rate; and open space above for installation of equipment. On the other hand, the slope in the insulated roof is dependent on the quality of the insulation.

Roof Overhang

Roof overhang can be used to shade the sidewalls of a building from direct and indirect solar radiation. However, the length of the roof overhang is dependent on the height of the sidewalls. Heat gain by the sidewall can be reduced to about 30% by roof overhang shading if properly applied at a roof slope of 45°.

Ridge Opening

Naturally, hot air rises above cooler air due to difference in air density. Introduction of ridge opening can aid ventilation through stack effect in the poultry house. Adequate setback between buildings is required to prevent inadequate airflow and circulation. However, ridge opening has been reported to be ineffective in insulated poultry houses because of temperature uniformity within the house.

Sidewall Openings

The sidewall consists of a dwarf wall built up to the roof eave with a permeable membrane such as a corrugated wire mesh and an adjustable curtain. A minimum height of 0.4 m is recommended to prevent the house from water seepage, direct and indirect solar radiation, pests and predators. The corrugated wire mesh allows easy airflow within and outside the building, while the adjustable curtain is used to control the flow and air velocity. However, the curtain may be transparent or of varying colors to aid its use in managing intermittent lighting scheme.

Building Obstruction

Adequate setback between buildings is required to prevent inadequate air exchange rates in building. Factors such as wind speed, wind direction and topography are major determinants for consideration in defining the optimal house spacing. However, the spacing between buildings can be determined by the expression below:

$$D = 0.4HL^{0.5}$$

where D, housing spacing (ridge of the closest wall of the next house); H, height of the adjacent building; L, length of the adjacent building.

Vegetation should be kept as minimal as possible and at average height to reduce the nest of wild birds and invasion of rodents and other predators. Also, the branch of trees should be kept at eaves level to prevent obstruction of airflow across the house.

Roof, End-wall and Sidewall Insulation

Farmers in the tropics have successfully used locally sourced materials such as thatched roof and bamboo as roofing materials for the construction of naturally ventilated poultry houses. However, a minimum R-value of 1.25 m^2 C/W was recommended for ceiling insulation in naturally ventilated poultry house. Environmental temperature higher than 40 °C would require a minimum R-value of 2.25 m^2 C/W. The various methods of insulating poultry house ceiling include dropped ceiling, rigid board insulation, spray polyurethane insulation and reflective insulation.

Cooling System

Rooftop sprinklers have proven to be efficient for substantially cooling the roof. However, material of choice in this situation must be able to withstand the constant exposure to water. Evaporative cooling in birds in hot weather can be subdued by using fogging system. With high water pressure it generates mist, which aids cooling in birds. However, the level of humidity within the house must be monitored for it could be detrimental to the health of birds at high temperature. Circulation fan eases heat stress by providing increased air velocity to increase convection cooling. Generally, circulation fans generate air velocity of 0.5 m/s or more and cover an area 15 times its horizontal diameter by five times its vertical diameter. Furthermore, for effective use of circulation fans it should be installed at the center 1–1.5 m above the floor and tilted downward at an angle 5°.

Vegetation

Shrubs and grasses reduce reflective and direct solar radiation by shading and convection cooling. Vegetation should be kept clean and trimmed to keep away predators and pests. The planting of tall trees along the sidewalls can provide a form of canopy to shade the sidewalls from exposure to direct or reflective solar radiation during the hot periods of the day.

Mechanically Ventilated Open Housing System

The limitation of attaining adequate internal environmental conditions required for optimum birds' performance under extreme weather conditions has led to the use of the mechanically ventilated housing system. Also, the mechanically ventilated house provides more control over air exchange, wind velocity and wind direction. Mechanically ventilated system entails the use of either positive or negative pressure system. The negative-pressure system which is the most commonly used in mechanical ventilated house, expels air out of the building by fans through an air inlet system to create low pressure within the house to allow fresh air to rush in through the same air inlet system.

The negative-pressure systems can be achieved through inlet or tunnel ventilation. Inlet ventilation system uniformly distributes exhaust fans and air inlets across the house whereas, tunnel ventilation exhausts fans are located at one end and inlet pipes at the other end. This provides the tunnel ventilation with an advantage of greater air speed in turn creating more positive air exchange.

House Construction

For proper ventilation control, it is required that the house be highly insulated and tightly constructed. However, the sidewall can be equipped with insulated adjustable curtains instead of solid wall for use in the cooler periods of the year or in case of power failure emergency. It is important to note that solid wall have higher insulation value that adjustable curtains.

Air Exchange

High external temperature coupled with the heat generated from the activities within the poultry house increases the temperature of the internal air. An effective mechanical ventilation system has to exchange the air quickly to ensure the internal air temperature maintains not more than 2.8 °C difference from the external air temperature. The expression below can be used to calculate the appropriate exhaust fan required for effective ventilation,

$$\text{Building surface heat}(\text{watts}) = (A/R) \times (T_o - T_i)$$

where, A, area of the building surface (m^2); R, insulation value of the wall material (m^2C/W); T_o, temperature outside (°C); T_i, temperature inside (°C).

The value of T_o is the hottest external temperature that is excepted of the external environment. However, when calculating heat gain for roof in a house with attic space, the value T_o it is assumed to be 55 °C whereas the T_o value for ceilings with insulation directly below the roof is assumed to be 65 °C. On the other hand, T_i is best assumed as 27 °C to ensure comfort for birds. The value of R will be the overall sum of the insulation value of the wall section. The total heat produced (sensible and latent) in commercial broiler is 7.9 W/kg while broiler, pullets and broiler breeders is 5.1 W/kg. The heat generated by birds is expressed below:

$$\text{Bird heat}(W) = \text{sensible heat}(W/kg) \times \text{weight of the bird}(kg)$$

where sensible heat, 50% of the total heat produced by birds,

$$\text{Total heat}(W) = \text{building surface heat}(W) + \text{bird heat}(W)$$

However, the air movement capacity to maintain 2.8 °C between intakes and exhaust air is expressed below:

$$\text{Aircapacity}(m^3/h) = \text{total heat}(W) \times 3.4 / 2.8^{\circ}C$$

Air Inlet System

There are a number of negative-pressure air inlet pipes used to control the internal climatic condition by controlling the entry location, speed and direction of fresh air. However, the exhaust fan determines how much air enters the house.

- Inlet speed: The pressure difference between the internal and external environment determines the entry speed of fresh air. However, the pressure is a function of the number and sizes of the air inlets. Therefore, the easy manipulation of differential pressure allows for possible control of airflow pattern within the building and of negative-pressure air inlet pipes used to control the internal climatic condition.
- Inlet area: For easy control and distribution of air within the poultry house, the exhaust fan must develop a static pressure of about 12–25 Pa.
- Air inlet control: Air inlet design should be located strategically as the direction of air depends on external climatic condition, age and class of the chicken. Normally, air inlets should be designed to direct air towards the ceiling at cooler time while another should be directed towards the floor during the hot periods of the year.
- Air inlet control: Static pressure of about 12–25 Pa was recommended for easy control of the air inlet for a static pressure above or below that range can lead to supply of insufficient air velocity.

Types of Inlet Ventilation System

- Cross ventilation: The exhaust fans are installed on one side while the air inlet pipes are along the other side of the poultry house. It is best suited to narrow poultry houses (less than 10 m) because it leads to difference in environmental conditions in the house with larger width.
- Sidewall inlet ventilation: The exhaust fans are placed below the air inlet pipes on both sides of the building walls. However, a distance not less than twice the diameter of the fan should be between the exhaust fans and the air inlet pipes. Air movement is directed towards the center, and drawn through the floor by the exhaust fans. It is also best suited for narrow house with not more than 12 m width.
- Attic inlet ventilation: The exhaust fans are placed on the lower sidewalls while, air inlets are placed in the ceiling. This kind of ventilation requires proper ceiling insulation and it best suitable for hot climate areas. The ventilation method is greatly recommended for rearing laying hen.

Air Movement Inlet Ventilated House

Fresh air enters through the air inlet pipes at a velocity of 3.5–6 m/s, however this velocity is quickly dropped to about 1 m/s depending on the size and type of the house.

Hence, circulation fans are used to boost the air speed to ensure air movement is sufficient in the building.

Tunnel Ventilation System

Tunnel ventilation system is designed to meet the specified air velocity and air exchange rate. However, the required air velocity is dependent on the class of birds in question. Table shows the recommended air speed required for rearing various classes of poultry birds.

Table: Recommended air velocities in tunnel-ventilated houses.

House type	Air speed (m/s)
Broilers	2.5–3
Pullets	1.75–2.25
Broiler breeders	2.25–3
Commercial layer	2.5–3

Tunnel Fan Capacity and Air Velocity

The tunnel fan capacity is determined by the same method used for inlet ventilation system. Unlike the inlet ventilation system where the adequate air velocity is propelled by circulation fan, the required average air velocity within tunnel house is calculated by the expression below:

$$\text{Tunnel fan capacity} = \text{desired air velocity} / (\text{cross - sectional area of the house}) \times 3600$$

where air velocity, m/s; tunnel fan capacity, m^3/h; cross section area, m^2.

However, it is important to note that the cross sectional area of the house adversely affect the air speed within the house. Therefore, it is advisable to design narrow and long house with lower ceilings. Consequently the expression below can be used to design the desired air velocity.

$$\text{Tunnel fan capacity} = \text{desired air velocity} / (\text{cross - sectional area of the house}) \times 3600$$

where desired air velocity, m/s; tunnel fan capacity, m^3/h; cross section area, m^2.

In cases where there is land constraints, air deflectors can be installed houses with large cross-sectional area to reduce the cross sectional area within the poultry house. Air deflectors are curtains that extend from the ceiling not more than 2.5–3 m from the ground. Air deflectors have been reported to increase air velocity for a distance approximately 1.2 and 6–9 m upward and downwind of the deflector respectively. However, it is important to ensure that the air deflector exceed 2.5 m from the ground to have it from disrupting the performance of fans and air exchange rate by increasing static pressure.

Air Velocity Distribution

Normally, the air velocity in a tunnel house is assumed uniform across the house. However, it can vary slightly depending on the smoothness of the building surfaces, presence of poultry equipment and other obstructions that deflect air. The difference between the air velocity in the center and the side of the house can vary from 15 to 40%.

Bi-directional Tunnel House

Generally, it is best to install the fans on one end and the inlet in opposite end to ensure the maximum air speed is achieved in the tunnel house. However in cases where the poultry house is over 180 m long and the air velocity required for airflow in one direction exceeds 3.5 m/s it is advisable to apply the bi-directional tunnel house system. The fans are located at end-walls of the building and the tunnel inlet at the center of the house. The air velocity in both direction is reduce to half of the required velocity while retaining the same air exchange rate to ensure the temperature difference between the inlet and the fan remains the same.

Tunnel Fan Placement

The fans can be installed at the end-walls or the sidewalls near the end, and this installation arrangement does not affect the performances of the fans. However, dead spot can be noticed when the fan is installed on the sidewalls as the width of the houses increase.

Tunnel Inlet Opening

In the absence of evaporative cooling pads, it is recommended that the inlet area should be at least 10% greater than the cross sectional area of the house. Meanwhile, the pad used determines inlet size for tunnel house with evaporative cooling pads. It is recommended that inlet opening on the sidewall should be installed as close as possible to the end wall. However, if the house width exceeds 15 m it is advisable to install the inlet openings on the end-wall.

Cool Weather Inlet System for Tunnel Ventilated Houses

It has been recommended that tunnel ventilated system should be used in hot weather because cool weather reduce the air exchange rate. Consequently, it was recommended that a minimum of 60% of the tunnel fan capacity should be controlled by the traditional inlet system before upgrading to tunnel ventilation for easy switch during cooler weathers.

Poultry Exhaust Fans

Types of Fans

- Exterior and interior shutter fans: It is the simplest type of exhaust fan. Its shutters are used to when the fan is not in use. However, the exterior shutter restricts airflow as air spins off its blades on contact. In the case of interior fan on the other hand, the shutters are on the intake side of the fan thus, lessening the restriction of airflow. It has bigger shutters, which allows for more air movement. Airflow is increased by 5–10% compared to exterior shutter fan.

- Discharge cone fans: It increases fan performance by 5–10% as it eases the transitioning of drawn towards the fans.
- Belt-drive fans: The fans blades are driven by a simple pulley mechanism. It may be upgraded with an automatic belt tensioner to prevent belt slippage.
- Direct-drive fans: The fan's blades are attached directly to the motor shaft eliminating the use of belt. They are less energy efficient compared to belt-driven fans.

Aquaculture

Seventy percent of the world's surface is covered in water, humans have realized its importance as a resource. For this reason, one of the areas heavily exploited regarding the use of water as a resource is aquaculture, especially in the production of food as opposed to using the terrestrial land. Aquaculture is the process of rearing, breeding and harvesting of aquatic species, both animals and plants, in controlled aquatic environments like the oceans, lakes, rivers, ponds and streams. It serves different purposes, including food production, restoration of threatened and endangered species populations, wild stock population enhancement, the building of aquariums, and fish cultures and habitat restoration.

Types of Aquaculture

There are different types of aquaculture:

- Depending on hydrobiological features.
- Depending on the motive of farming.
- Depending on special operational techniques.

Mariculture

Mariculture is aquaculture that involves the use of seawater. It can either be done next to an ocean, with a sectioned off part of the ocean or in ponds separate from the ocean, but containing seawater all the same. The organisms bred here range from molluscs to seafood options like prawn and other shellfish, and even seaweed.

Growing plants like seaweed are also part of mariculture. These sea plant and animal species find many uses in manufacturing industries such as in cosmetics and jewellery where collagen from seaweed is used to make facial creams. Pearls are picked from molluscs and made into fashion items.

Fish Farming

Fish farming is the most common type of aquaculture. It involves the selective breeding of fish, either in freshwater or seawater, with the purpose of producing a food source for consumption. Fish farming is highly exploited as it allows for the production of a cheap source of protein.

Furthermore, fish farming is easier to do than other kinds of farming as fish are not care-intensive, but only requiring food and proper water conditions as well as temperatures. The process is also less land-intensive as the size of ponds required to grow some fish species such as tilapia is much smaller than the space required growing the same amount of protein from beef cattle.

Algaculture

Algaculture is a type of aquaculture involving the cultivation of algae. Algae are microbial organisms that share animal and plant characteristics. They are sometimes motile like other microbes, but they also contain chloroplasts that make them green and allow them to photosynthesize just like green plants. However, for economic feasibility, they have to be grown and harvested in large numbers. Algae are finding many applications in today's markets.

Integrated Multi-trophic Aquaculture (IMTA)

IMTA is an advanced system of aquaculture where different trophic levels are mixed into the system to provide different nutritional needs for each other. Notably, it is an efficient system because it tries to emulate the ecological system that exists in the natural habitat. The IMTA makes use of this inter-tropical transfer of resources to ensure maximum resource utilization by using the waste of larger organisms as food sources for the smaller ones. The practice ensures the nutrients are recycled, meaning the process is less wasteful and produces more products.

Inland Pond Culture

This usually involves inland artificial ponds of about 20 acres in size and about 6-8ft deep. It is common to see aeration systems connected to the pond, to introduce air into the ponds. This enhances the supply of oxygen and also reduces ice formation in the winter season. In China, over 75% of the farmed freshwater fish are produced in constructed ponds, and nearly all of the farmed catfish are raised in ponds in the U.S.

Recirculating Systems

This involves a closed set of chambers (units) where fish is kept in one and water treatment kept in another. It is highly dependent on the power supply, as water has to be pumped constantly through the fish chambers. As water flows through the treatment chamber, particulate matter is filtered out and air introduced. This closed system controls the salinity, temperature, oxygen and anything that can cause harm to the fish. It is an environmentally friendly system because very little new water is introduced to replace water that evaporated. The residue from the filters is also disposed of in a responsible manner.

Open-net Pen and Cage Systems

Open-net pen and Cage systems are often found offshore and in freshwater lakes. Mesh cages of between 6 and 60 cubic feet (pens) are installed in the water with the fish inside it. With a high concentration of fish in the pens, waste, chemicals, parasites and diseases are often exchanged in the immediate water environments. The fish also attract predatory animals (bigger fish), which are often entangled in the nets. This system uses public water; therefore, environmental regulation and some authorization protocols must be respected.

Flow-through/Raceway

This is a system made of long units stocked with fish. The units have feeding stations attached to them. Water is diverted from flowing water and fed into the raceway units flowing downstream. Down the end of the unit, waste is collected and disposed of. Raceways are common for culturing trout.

Benefits of Aquaculture

Economic Benefits

- Alternative Food Source: Fish and other seafood are good sources of protein. They also have more nutritional value like the addition of natural oils into the diet, such as omega 3 fatty acids. Also, since it offers white meat, it is better for the blood to reduce cholesterol levels as opposed to beef's red meat. Fish is also easier to keep compared to other meat-producing animals as they are able to convert more feed into protein. Therefore, its overall conversion of a pound of food to a pound of protein makes it cheaper to rear fish as they use the food more efficiently.

- Alternative Fuel Source: Algae are slowly being developed into alternative fuel sources by having them produce fuels that can replace contemporary fossil fuels. Algae produce lipids that, if harvested, can be burned as an alternative fuel source whose only by-products would be water when burnt.

- Such a breakthrough could ease the dependency of the world on drilled fossil fuels as well as reduce the price of energy by having it grown instead of drilling petroleum. Moreover, algae fuel is a cleaner and farmable source of energy, which means it can revolutionize the energy sector and create a more stable economy that avoids the boom-bust nature of oil and replaces it with a more abundant fuel source.

- Increase Jobs in the Market: Aquaculture increases the number of possible jobs in the market. It provides both new products for a market and creates job opportunities as labor is required to maintain the pools and harvest the organisms grown. The increase in jobs is mostly realized in third world countries as aquaculture provides both a food source and an extra source of income to supplement those who live in these regions.

- Aquaculture also saves fishermen time as they do not have to spend their days at sea fishing. It allows them free time to pursue other economic activities like engaging in alternative businesses. This boosts entrepreneurship and provides more hiring possibilities and more jobs.

- Reduce Sea Food Trade Deficit: The seafood trade in America is mainly based on trade from Asia and Europe, with most of it being imported. The resultant balance places a trade deficit on the nation. Aquaculture would provide a means for the reduction of this deficit at a lower opportunity cost as local production would mean that the seafood would be fresher. It would also be cheaper due to reduced transport costs.

Environmental Benefits

Creates a Barrier against Pollution with Mollusc and Seaweed

Molluscs are filter feeders, while seaweed acts a lot like the grass of the sea. Both these organisms sift the water that flows through them as brought in by the current and clean the water. This provides a buffer region that protects the rest of the sea from pollution from the land, specifically from activities that disturb the sea bed and raise dust. Also, the economic benefits of molluscs and seaweed can create more pressure from governments to protect their habitats as they serve economic importance. The financial benefits realized provide an incentive for the government to protect the seas in order to protect seafood revenue.

Reduces Fishing Pressure on Wild Stock

The practice of aquaculture allows for alternative sources of food instead of fishing the same species in their natural habitats. Population numbers of some wild stocks of some species are in danger of being depleted due to overfishing and uncontrolled exploitation. The use of unsustainable fishing methods such as bottom trawlers is also reduced.

Aquaculture provides an alternative by allowing farmers to breed those same species in captivity and allow the wild populations to revitalize. The incentive of less labor for more gains pushes fishers to convert to fish farmers and make even more profit than before. It also allows the control of the supply of the fish in the market, giving them the ability to create surplus stock or reduce their production to reap the best profits available.

Low Environmental Impact

Aquaculture poses a low risk to the environment. The impact is mostly local and temporary. In some cases, aquaculture can benefit the environment. Where filter-feeding shellfish, such as oysters, are cultured in-situ, water quality in ponds and lakes can improve.

Fish and shellfish can also be farmed using methods that do not harm the environment, and that helps meets the growing demand for seafood by supplementing wild harvests. Especially for offshore systems, the bio-security systems, cameras and surveillance infrastructure, as well as trained inspectors, ensure that farms are complying with environmentally safe practices. This helps to reduce diseases transfer in the waters and so on.

Water Usage

Aquaculture systems often take advantage of harvested runoffs, storm water and surface water. This reduces the dependency on other sources of water supply. In addition to this, ponds maintain soil moisture in their vicinity, thereby conserving natural resources.

Importance of Aquaculture

- Health Benefit: All over the world, the demand for seafood has increased because people have learned that seafood are healthier and help fight cardiovascular disease, cancer, alzheimer's and many other major illnesses. Now seafood has become part of regular diets.
- Sustainable use of Sea Resources: Aquaculture provides alternatives for fishing from the sea. An increase in demand for food sources and globalization has led to an increase in fishing. Aquaculture is currently estimated to account for approximately 13 percent (10.2 million tons) of world fish production. Yet, this has led fishermen to become selfish and overfish the desired or high-demand species. Through aquaculture, it provides both an alternative and opportunity for wild stocks to replenish over time.
- Conservation of Biodiversity: Aquacultures also protect biodiversity by reducing the fishing activities on the wild stock in their ecosystems. By providing alternatives to fishing, there is a reduced attack on the wild populations of the various species in the sea. Reduced action of fishing saves the diversity of the aquatic ecosystem from extinction due to overfishing.
- Increased Efficiency, More Resources for Less Effort: Fish convert feed into body protein more efficiently than cattle or chicken production. It is much more efficient, meaning that the fish companies make more food for less feed. Such efficiency means that less food and energy is used to produce food, meaning that the production process is cheaper as well. It saves resources and even allows for more food to be produced, leading to secure reserves and less stress on the environment. Aquacultures will add to wild seafood and make it cheaper and accessible to all, especially in regions where they depend on imported seafood products.
- Reduced Environmental Disturbance: By increasing aquaculture, fish farming in specific, there is a reduced need for the fishing of the wild stock. As an outcome, it puts less stress on the ecosystem and equally reduces human interference. Actions of motorboats and other human influences such as the removal of viable breeding adult fish are all stresses put on the aquatic ecosystems, and their discontinuation allows the ecosystem to flourish and find their natural balance.

Beekeeping

Beekeeping is the care and management of colonies of honeybees. They are kept for their honey and other products or their services as pollinators of fruit and vegetable blossoms or as a hobby. The practice is widespread: honeybees are kept in large cities and villages, on farms and rangelands, in forests and deserts, from the Arctic and Antarctic to the Equator. Honeybees are not domesticated. Those living in a man-made domicile called a beehive or hive are no different from those living in a colony in a tree.

In antiquity people knew that bees produce delicious honey, that they sting, and that they increase their numbers by swarming. By the 17th century they had learned the value of smoke in controlling them and had developed the screen veil as protection against stings. From the 17th to the 19th century, the key discoveries upon which modern beekeeping are founded were made. These included the mystery of the queen bee as the mother of nearly all the occupants of the hive, her curious mating technique, parthenogenetic development, the movable frame hives, and the fact that bees rear a new queen if the old one disappears.

Given this knowledge, people were able to divide a colony instead of relying on natural swarming. Then the development of the wax-comb foundation, the starter comb on which bees build straight, easily handled combs, and the discovery that honey can be centrifuged or extracted from them and the combs reused, paved the way for large-scale honey production and modern commercial beekeeping. The identification of bee diseases and their control with drugs, the value of pollen and pollen substitutes in producing strong colonies, and the artificial insemination of queens have increased the honey-production efficiency of colonies.

Honeybees and their Colonies

Honeybees

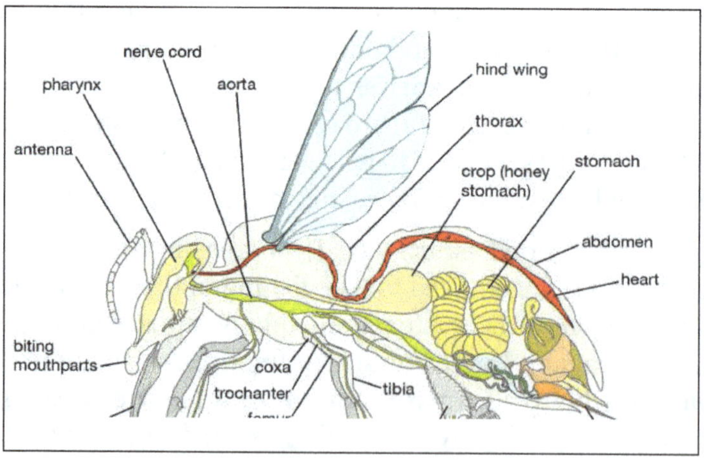

Body plan of a honeybee.

Honeybees belong to the order Hymenoptera and to one of the Apis species. Honeybees are social insects noted for providing their nests with large amounts of honey. A colony of honeybees is a highly complex cluster of individuals that functions virtually as a single organism. It usually consists of the queen bee, a fertilized female capable of laying a thousand or more eggs per day; from a few to 60,000 sexually undeveloped females, the worker bees; and from none to 1,000 male bees, or drones. The female of most species of bees is equipped with a venomous sting.

Honeybees collect nectar, a sugary solution, from nectaries in blossoms and sometimes from nectaries on the leaves or stems of plants. Nectar may consist of 50 to 80 percent water, but when the bees convert it into honey it will contain only about 16 to 18 percent water. Sometimes they collect honeydew, an exudate from certain plant-sucking insects, and store it as honey. The primary carbohydrate diet of bees is honey. They also collect pollen, the dust like male element, from the anthers of flowers. Pollen provides the essential proteins necessary for the rearing of young bees. In the act of collecting nectar and pollen to provision the nest, the bees pollinate the flowers they visit. Honeybees also collect propolis, a resinous material from buds of trees, for sealing cracks in the hive or for covering foreign objects in the hive that they cannot remove. They collect water to air-condition the hive and to dilute the honey when they consume it. A populous colony in a desirable location may, in a year's time, collect and carry into the hive as much as 1,000 pounds (450 kilograms) of nectar, water, and pollen.

Nectar: A honeybee (Apis) drinking nectar from a flower.

Bees secrete beeswax in tiny flakes on the underside of the abdomen and mold it into honeycomb, thin-walled, back-to-back, six-sided cells. The use of the cell varies depending on the needs of the colony. Honey or pollen may be stored in some cells, while the queen lays eggs, normally one per cell, in others. The area where the bees develop from the eggs is called the broodnest. Generally, honey is stored toward the top of the combs and pollen in cells around the broodnest below the honey.

Bees on a honeycomb.

The bees maintain a uniform temperature of about 93 °F (34 °C) in the broodnest regardless of outside temperature. The colony can survive daily maximum temperatures of 120 °F (49 °C) if water is available with which they can air-condition the cluster. When the temperature falls below about 57 °F (14 °C), the bees cease flying, form a tight cluster to conserve heat, and await the return of warm weather. They can survive for several weeks in temperatures of −50 °F (−46 °C). When summer flowers bloom in profusion, the queen's egg-laying is stimulated, the cluster expands, and honey accumulates in the combs. When the large number of young bees emerges, the domicile becomes crowded.

Swarming

When the colony becomes crowded with adult bees and there are insufficient cells in which the queen can lay large numbers of eggs, the worker bees select a dozen or so tiny larvae that would otherwise develop into worker bees. These larvae are fed copiously with royal jelly, a whitish food with the consistency of mayonnaise, produced by certain brood-food glands in the heads of the worker bees. The cell in which the larva is developing is drawn out downward and enlarged to permit development of the queen. Shortly before these virgin queens emerge as adults from their queen cells, the mother queen departs from the beehive with the swarm. Swarming usually occurs during the middle of a warm day, when the queen and a portion of the worker bees (usually from 5,000 to 25,000) suddenly swirl out of the hive and into the air. After a few minutes' flight, the queen alights, preferably on a branch of a tree but sometimes on a roof, a parked automobile, or even a fire hydrant. All the bees settle into a tight cluster around her while a handful of scouts reconnoitre a new home site. When the scout bees have located a new domicile, the cluster breaks. The swarm takes to the air and in a swirling mass proceeds to the new home. Swarming is the bees' natural method of propagation or increase.

Queen Bee

Back in the parent colony, the first queen to emerge after the mother queen departs with the swarm immediately attempts to destroy the others. If two or more emerge at the same time, they fight to the death. When the surviving virgin is about a week old, she soars off on her mating flight. To maintain genetic diversity within a colony, a queen frequently mates with more than one drone (called polyandry) while in the air. She may repeat the mating flights for two or three successive days, after which she begins egg laying. She rarely ever leaves the hive again except with a swarm. Normally, sufficient sperm are stored in her sperm pouch, or spermatheca, to fertilize all the eggs she will lay for the rest of her life. The drones die in the act of mating.

The queen can live up to five years, although many beekeepers replace the queen every year or two. If she is accidentally killed or begins to falter in her egg-laying efficiency, the worker bees will rear a "supersedure" queen that will mate and begin egg laying without a swarm emerging. She ignores the mother queen, who soon disappears from the colony.

Worker Bees

Worker bees live about six weeks during the active season but may live for several months if they emerge as adults in the fall and spend the winter in the cluster. As the name implies, worker bees do all the work of the hive, except the egg laying.

Drones

Drones are reared only when the colony is populous and there are plentiful sources of nectar and pollen. They usually live a few weeks, but they are driven from the hive to perish when fall or an extended period of adversity comes upon the colony. The only duty of the drone is to mate with the queen.

The queen can lay drone (unfertilized) eggs in the drone cells. If she is not allowed to mate or if her supply of sperm is exhausted, she will lay unfertilized eggs in worker cells. The development of unfertilized eggs into adult drones is known as parthenogenesis. Occasionally a colony may become queenless and unable to develop another queen. Then some of the worker bees begin to lay eggs, often several to a cell, and these develop into drones. A colony that has developed laying workers is difficult to requeen with a laying queen.

Colony Manipulation

The Yearly Work Cycle

The beekeeper's year starts in early fall. At that time he requeens the colonies whose queens are not producing adequate amounts of brood and makes sure that each colony has sufficient stores: at least 50 pounds (22 kilograms) of honey and several frames filled with pollen. Some beekeepers also feed the drug fumagillin to reduce possible damage to the adult bees by nosema disease. The colonies need a sunny exposure and protection from cold winds. Some beekeepers in northern and mountainous areas wrap their colonies with insulating material in winter. A few beekeepers kill their bees in the fall, harvest the honey, store the empty equipment, then restock with a two- or three-pound (0.8- or 1.4-kilogram) package of bees and a young queen the following spring.

Beekeeping: A beekeeper collecting a bee swarm.

If the colonies are well prepared in the fall, they need little attention during the winter. But in early spring an examination of the colonies by the beekeeper is important. Frequently, strong colonies exhaust their food supply and starve only a few days before flowers begin to bloom in abundance. Only a few pounds of sugar syrup, 50-50 sugar water, or a honey-filled comb from another more prosperous colony might save such a starving colony. Again fumagillin may be fed to the colony, and some beekeepers also feed a cake of pollen substitute or pollen supplement. Honey is not fed to the colonies unless the beekeeper is sure about its source. Honey from colonies affected by the brood disease American foulbrood could infect his colonies and cause a serious loss.

As the spring season advances, the cluster size increases from the low population of 10,000 to 20,000 bees that survived the winter. To accommodate the increased size of the cluster and broodnest, the keeper adds more supers, or boxes of combs. If the combs are so manipulated that the queen can continually expand her egg-laying area upward, the colony is unlikely to swarm. This can be achieved by placing empty combs or combs in which brood is about ready to emerge at the top of the cluster and combs filled with eggs or young brood toward the lower part of the broodnest. The beekeeper wants the colony to reach its peak of population, 50,000 to 60,000 bees, at the beginning of the major nectar flow.

The bees in a swarm, having departed the hive with a full stomach of honey, rarely sting. The usual way to capture them is to place a hive or upturned box beneath or nearby, then shake or smoke the bees to force the queen and a majority of the bees into it. The others follow. After the swarm is safely inside the box, it can be removed to a permanent location. Regulations governing the keeping of bees usually require the bees to be kept in hives with movable combs. If the bees are captured in a box, they are generally transferred into a movable-frame hive within a few days so the new honey and comb will not be lost in the transfer.

Requeening a Colony

When a beekeeper requeens a colony, he removes the failing or otherwise undesirable queen and places a new one in a screen cage in the broodnest. After a few days the colony becomes adjusted to her and she can be released from the cage. A strange queen placed in the cluster without this temporary protection usually will be killed at once by the workers. Queens usually are shipped in individual cages of about three cubic inches (50 cubic centimetres) with about half a dozen attendant bees and a ball of specially prepared sugar candy plugging one end of the cage. When the cage is placed in the hive, the bees from both sides eat the candy. By the time the candy is consumed and the bees reach each other, their odours have become indistinguishable, and the queen emerges from the cage into the colony and begins her egg-laying duties.

Beekeeping Equipment

Standard tools of the beekeeper are: the smoker to quell the bees; a veil to protect the face; gloves for the novice or the person sensitive to stings; a blunt steel blade called a hive tool, for separating the frames and other hive parts for examination; the uncapping knife, for opening the cells of honey; and the extractor, for centrifuging the honey from the cells.

Bee Stings

The worker bee sting is barbed, and in the act of stinging it is torn from the bee. It has a venom-filled poison sac and muscles attached that continue to work the sting deeper into the flesh for several minutes and increase the amount of venom injected. To prevent this, the sting should be scraped loose (rather than grasped and pulled out) at once. Bee stings are painful, and no one becomes immune to the pain. Immunity to the swelling is usually built up after a few stings.

However, normal reaction to a bee sting is immediate, intense pain at the site of the sting. This lasts for a minute or two and is followed by a reddening, which may spread an inch or more. Swelling may not become apparent until the following day. Occasionally, acute allergic reactions develop from a sting, usually with persons who have other allergic problems. Such a reaction becomes evident in less than an hour and may consist of extreme difficulty in breathing, heart irregularity, shock, splotched skin, and speech difficulty. Such persons should obtain the services of a medical doctor immediately.

Bee Products

- Honey production: Honey is marketed in several different forms: liquid honey, comb honey, and creamed honey. Sometimes the predominant floral type from which the honey was collected is indicated.
- Liquid honey: If liquid (strained, extracted) honey is desired, additional supers are added directly above the brood nest. When one is largely filled, it is raised and another is placed underneath. This may continue until several have been filled, each holding from 30 to 50 pounds (14 to 23 kilograms), or until the nectar flow has ended. After the bees have evaporated the water until the honey is of the desired consistency and sealed in the cells, the combs are removed, the cells uncapped with the uncapping knife, and the honey extracted. The removed honey is immediately heated to about 140 °F (60 °C), which thins it and destroys yeasts that can cause fermentation. It is then strained of wax particles and pollen grains, cooled rapidly, and packaged for market.
- Comb honey: In production of honey in the comb, or comb honey, extreme care is necessary to prevent the bees' swarming. The colony must be strong, and the bees must be crowded into the smallest space they will tolerate without swarming. New frames or sections of a frame with extra-thin foundation wax, added at exactly the right time for the bees to fill without destroying them, are placed directly above the brood nest. The bees must fill and seal the new comb to permit removal within a few days, or it will be of inferior quality. As rapidly as sections

are removed, new sections are added, until the nectar flow subsides. Then these are removed and the colony given combs to store its honey for the winter.

- Creamed honey: Almost all honey will granulate or turn to sugar. Such honey can be liquefied without materially affecting its quality by placing the container in water heated to about 150 °F (66 °C). Liquid and granulated honey is sometimes blended, homogenized, and held at a cool temperature, which speeds uniformly fine granulation. If properly processed, the granules will be extremely fine; the honey, which has a smooth, creamy appearance, is referred to as creamed honey.

Floral Types

Some honeys are sold by floral type; that is, they are given the name of the predominant flowers visited by the bees when they accumulated the honey. The beekeeper has no way to direct the bees to a particular source of food but through experience learns which plants are the major sources of honey. Different flowers produce different colours and flavours of honey. It may be heavy-bodied or thin-bodied, dark or light, mild-flavoured or strong-flavoured. Most honey has been blended by the beekeeper to a standard grade that can be supplied and marketed year after year.

Beeswax

Beeswax is a by-product of beekeeping in most areas. When beekeepers uncap or break honeycombs or have unusable combs, they try to salvage the beeswax. First, they recover as much honey from the combs as possible by drainage or extraction. Then they place the material in water heated to slightly over 145 °F (63 °C). This melts the wax, which rises to the surface. After it cools and hardens, the cake of wax is removed and refined for reuse in comb foundation. Beeswax has many other uses: in quality candles, cosmetics, agriculture, art, and industry. In some areas bees are manipulated primarily for wax production. Wax is a highly stable commodity that can be transported long distances under unfavourable conditions without damage.

Bees Reared for Sale

Queens are reared for sale to other beekeepers for requeening established colonies or for adding to a 2- or 3-pound (0.9- or 1.4-kilogram) package of 8,000 to 10,000 live bees to form new colonies or replenish weak ones. The queens are produced when the beekeeper cages the reigning queen in a colony, then inserts into the cluster from 30 to 60 queen cell bases into which young (one-day-old) worker larvae have been transferred. Queens can be artificially inseminated with sperm from drones of a known source, but most beekeepers let the queens mate naturally. The live bees are shaken from the combs of the colony through a funnel into screen-wire cages.

Pollination

The greatest value of bees is in their service as pollinators. Some 90 crops grown in the United States alone are dependent on insect pollination, performed primarily by the honeybee. The average colony of bees is worth from 20 to 40 times as much in the pollination of crops as it is in the production of honey. The value of bees in the pollination of ornamental plants has never been calculated. Bees are also valuable in the pollination of some forest and range plants that produce seeds on which birds and other wildlife feed.

Bee approaching a flower.

When bees are used in the pollination of crops, the beekeeper places the colonies within or adjacent to the field to be pollinated. The majority of the roughly 1,000,000 colonies that are used for pollination are used in alfalfa-seed fields and almond and apple orchards. The colonies are distributed at the rate of two or more per acre in groups every 0.1 mile (0.16 kilometre) throughout alfalfa fields. Two colonies per acre are recommended for almond orchards and about one colony per acre in apple orchards.

Some growers prefer to have the colonies placed alongside the orchard; others want them distributed in small groups within the orchard. Bees also are used regularly by growers of many other crops: blueberries, cantaloupes, cherries, clovers, cucumbers, cranberries, cutflower seed, plums and prunes, vetch, and watermelon.

Disease and Pest Control

Honeybees have diseases and enemies: diseases of the brood; diseases that affect only the adult bees; insect enemies of the adults and of the comb; and other enemies, including toads, lizards, birds, mice, skunks, and bears.

Diseases

American foulbrood, caused by a spore-forming bacterium, Bacillus larvae, is the most serious brood disease. It occurs throughout the world wherever bees are kept and affects workers, drones, and queens. The spores are highly resistant to heat and chemicals.

A comb containing brood severely infected with this disease has a mottled appearance caused by the mixture of healthy capped brood interspersed with diseased or empty cells formerly occupied by diseased brood. The decayed mass has a typical ropiness when dug into, which is one of its identifying characteristics.

American foulbrood can be spread to healthy colonies by transferring equipment or allowing the bees to feed on honey from infected colonies. Sulfathiazole and Terramycin are widely used to control the disease. Many countries and most states in the U.S. require the destruction by fire of diseased colonies and have apiary inspectors to enforce the regulations. European foulbrood is caused by a nonsporeforming bacterium, Streptococcus pluton, but Bacillus alvie and Acromobacter eurydice are often associated with Streptococcus pluton. This disease is similar in appearance to American foulbrood. In some instances it severely affects the colonies, but they recover so that colony destruction is not necessary. Terramycin can control the disease.

Sacbrood is caused by a virus and is superficially similar to the foulbrood diseases. It can appear and disappear spontaneously but is seldom serious. No chemical control is needed. If the problem persists, the beekeeper usually requeens the colony.

Chalk brood is caused by the fungus Ascosphaera apis. The larvae victims of this disease have a chalky white appearance. Stonebrood, which affects both brood and adults, is also caused by a fungus, Aspergillus flavus, which can usually be isolated from bees that have stonebrood.

Nosema disease, caused by the microsporidian Nosema apis, is the most serious disease of adult bees. It is widespread, causes heavy losses in honey production, and severely weakens colonies. The external symptoms of bees with nosema disease are not apparent. The disease is transmitted from adult to adult by ingestion of the spores that soon germinate in the ventriculus, or main, stomach. An infected ventriculus is normally swollen, soft, and grayish white. A degree of control may be obtained by feeding the colony the drug fumagillin.

Acarine disease is caused by the mite Acarapis woodi that gets into the tracheae of the bee through its breathing holes or spiracles in its thorax or midsection. Bees affected by this mite are unable to fly, have disjointed wings and distended abdomens. There is presently no good control for this mite. The only U.S. federal law pertaining to bees was passed to prevent the importation of adult bees carrying this mite into the United States. Two other mites, Varroa destructor and Tropilaelaps clareae, which are native to Asia, are serious problems for beekeepers. V. destructor is now commonly found in Europe and North America, where it is capable of devastating entire colonies of honeybees. There are other minor diseases of adult bees, but they seldom cause serious problems.

Pests

The greater wax moth, Galleria mellonella, is a lepidopterous insect that, in its larval stage, destroys combs. It does not attack adult bees but may begin destruction of combs of a weak colony long before the bees are gone. It can also destroy stored combs of honey. When the larvae are ready to pupate, they often eat out a place to spin their cocoons in the soft wood of the beehive, damaging frames and other hive parts. The best control for this pest is keeping colonies strong. Stored combs are fumigated, kept in a cold room, or stacked in such a way that a strong air draft flows around them.

The larvae of the lesser wax moth, Achroia grisella, and cause damage to stored combs similar to that of the greater wax moth. The Mediterranean flour moth larva, Anagasta kuehniella, feeds on pollen in the combs and causes some damage. Control for both of these moths is the same as for the greater wax moth. The bee louse, Braula caeca, is a tiny, wingless member of the fly family that is occasionally found on bees. It feeds on nectar or honey from the mouthparts of its host. Its larvae burrow in the cappings of honey combs.

Ants sometimes invade hives and disrupt or kill the bees. Termites can damage or destroy hive parts placed on the soil. Other insects, such as dragonflies (Odonata), robberflies (Diptera), praying mantises (Orthoptera), ambush bugs (Hemiptera), and certain wasps and yellow jackets (Hymenoptera) are natural enemies of the honeybee.

Predators

Mice frequently enter the hive in winter when the bees are clustered, or they get into stored combs and despoil or damage them by chewing the frames and combs to construct their nest. Skunks devour large numbers of bees at the hive entrance, usually at night. Fences, traps, and poison are used against them. Bears eat the honeybees and the brood in the hive, usually destroying it and its contents in the process. In bear country, electric fences and traps are used to protect bee colonies.

At times bees become their own deadly enemy. If honey is exposed to them when no flowers are in bloom and the weather is mild, the bees from different colonies will fight over it. Sometimes this fighting, or robbing, becomes intense and spreads from hive to hive in moblike action. If all the bees in one colony are killed, the honey is quickly stolen and carried into other hives. This further intensifies the robbing so that a cluster that was carrying honey into its hive a few minutes earlier is attacked, all of its occupants killed, the honey again stolen, and the process repeated. Usually, once robbing becomes intense, only darkness or foul weather will stop it.

Colony Collapse Disorder

One of the most mysterious disorders to strike honeybee colonies in the modern era is colony collapse disorder (CCD). It is characterized by sudden colony death, with a lack of healthy adult bees inside the hive. While the underlying cause is not known, it appears that the disorder affects the adult bees' ability to navigate. They leave the hive to find pollen and never return. Honey and pollen are usually present in the hive, and there is often evidence of recent brood rearing. In some cases the queen and a small number of survivor bees may remain in the brood nest. CCD is also characterized by delayed robbing of the honey in the dead colonies by other, healthy bee colonies in the immediate area, as well as slower than normal invasion by common pests, such as wax moths and small hive beetles. The disorder appears to affect only the European honeybee (Apis mellifera).

CCD was first reported in autumn 2006 by a commercial beekeeper in Pennsylvania, who had colony losses estimated at 80 to 90 percent. Colony losses continued to be reported by other beekeepers in 35 states throughout the United States during the spring and summer of 2007, with many beekeepers losing anywhere from 30 to 90 percent of their hives. Beekeepers in other countries, including Canada, Portugal, Italy, Spain, Greece, Germany, Poland, France, and Switzerland, also reported substantial losses of honeybees. In the following years, the syndrome continued to impact honeybee colonies, though the percentage of colonies lost annually appeared to decline. Nonetheless, the potential economic impact on agriculture is great; annually in the United States alone an estimated $15 billion of crops are pollinated by honeybees.

Studies of adult honeybee carcasses from affected colonies indicate that the bees are infected with a number of pathogens and parasites, including viruses, species of Nosema, and the phorid fly Apocephalus borealis. However, scientists have not reached a definitive conclusion on whether a single pathogen is the root cause of the disorder, and many scientists suspect that a combination of factors are involved, such as a weakened immune system, brought on by colony stress, and the presence of pathogens, which are a constant threat and can be numerous in honeybee colonies. In addition, pesticides such as neonicotinoids (insecticides based on derivatives of nicotine) can be toxic to honeybees and are suspected of causing or contributing to CCD.

Sericulture

Sericulture is an agro-based industry. It involves rearing of silkworms for the production of raw silk, which is the yarn obtained out of cocoons spun by certain species of insects. Cultivation to feed the silkworms that spin silk cocoons and reeling the cocoons to unwind the silk filament for value added advantages like process and weaving are the major activities of sericulture.

Advantages of Sericulture

High Employment Potential

- 60 lakh persons are engaged in various sericulture activities in the country.
- It is calculable that Sericulture can generate employment at 11 man days per kg of raw silk production (in on-farm and off-farm activities) throughout the year. This potential ispar-excellence and no other business generates this type of employment, especially in rural areas, hence, sericulture is used as a tool for rural reconstruction.

Eco-friendly Activity

- As a perennial crop with good foliage and root-spread, mulberry contributes to conservation and provides green cover.
- Waste from silkworm rearing will be recycled as inputs to garden.
- Dried mulberry twigs and branches are used as fuel in place of firewood and therefore reduce the pressure on vegetation/forest.
- Being a labour intensive and predominantly agro-based activity, involvement of smoke-emitting machinery is minimal.
- Developmental programmes initiated for mulberry plantation are mainly in up-land areas where un-used cultivable land is made productive.
- Intercropping can also be done with numerous plantations.
- Being a deep-rooted perennial plant it can be raised in vacant lands, hill slopes and watershed areas.
- Currently, only concerning 0.1 % of the cultivable land in the country is beneath mulberry cultivation.

Women Friendly Occupation

Women constitutes over hour of those employed in down-stream activities of sericulture in the country. This is the result of sericulture activities ranging from mulberry garden management, leaf harvesting and silkworm rearing is more effectively taken up by the women folks. Even silk reeling business together with weaving is largely supported by them.

References

- Facts-about-cattle-farming: livestocking.net, Retrieved 07, August 2020
- Sheep-farming: roysfarm.com, Retrieved 11, March 2020
- Goat-system-productions-advantages-and-disadvantages-to-the-animal-environment-and-farmer, goat-science: intechopen.com, Retrieved 02, July 2020
- Horse-farm-management: hobbyfarms.com, Retrieved 06, January 2020
- Nutrients-Cycle-within-Swine-Production-Generation-Characteristics-Treatment-and-Revaluation-341741033: researchgate.net, Retrieved 21, June 2020
- Poultry-farming: sentientmedia.org, Retrieved 15, February 2020
- List-of-poultry-farm-equipment: agricfarming.com, Retrieved 29, May 2020
- Poultry-housing-and-management, poultry-an-advanced-learning: intechopen.com, Retrieved 04, April 2020
- Aquaculture-types-benefits-importance: conserve-energy-future.com, Retrieved 19, July 2020
- Sericulture-an-introduction-to-silk-cultivation-and-production-in-india-along-with-its-policy-initiatives, agripedia: krishijagran.com, Retrieved 27, January 2020

4

Techniques of Animal Breeding

In animal breeding, animals with desirable genetic traits are selected and bred to maintain these traits in future generations. The process of producing offspring by mating of genetically related individuals is known as inbreeding. Outbreeding refers to the matings between individuals who have no common ancestors, or are distantly related. This chapter has been carefully written to provide an easy understanding of the varied facets of animal breeding.

A group of animals related by descent and similar in most characters like general appearance, features, size, configuration, etc. are said to belong to a 'breed'.

Animal breeding is producing improved breeds of domesticated animals by improving their genotypes through selective mating.

Objectives of Animal Breeding

The main objectives of animal breeding are:

(i) Improved growth rate,

(ii) Increased production of milk, meat, egg, wool, etc.,

(iii) Superior quality of milk, meat, eggs, wool, etc.,

(iv) Improved resistance to various diseases,

(v) Increased productive life, and

(vi) Increased or, at least, acceptable reproduction rate.

Methods of Animal Breeding

Two methods of animals breeding are: inbreeding and out breeding, based mainly on breeding work with cattle.

1. Inbreeding

When breeding is between animals of the same breed for 4-6 generations, it is called inbreeding. Inbreeding may be explained by taking an example of cows and bulls. Superior cows and superior bulls of the same breed are identified and mated. The progeny obtained from such mating are evaluated and superior males and females are identified for further mating. A superior female, in the case of cattle, is the cow that produces more milk per lactation.

On the other hand, a superior male is that bull, which gives rise to superior progeny as compared to those of other males. As the homozygous pure lines developed by Mendel as described in Chapter 5, a similar strategy is used for developing pure lines in cattle as was used in case of peas. Inbreeding, as a rule, increases homozygosis.

Thus inbreeding is necessary if we want to develop a pure line in any animal. Inbreeding exposes harmful recessive genes that are eliminated by selection. It also helps in accumulation of superior genes and elimination of less desirable genes. But continued inbreeding reduces fertility and even productivity.

This is called inbreeding depression. In this condition, the selected animals of the breeding population should be mated with superior animals of the same breed but unrelated to the breeding population. This often helps in restoring fertility and yield.

2. Out breeding

Out breeding is the breeding between the unrelated animals which may be between individuals of the same breed (but having no common ancestors) or between different breeds (cross breeding) or different species (interspecific hybridization).

(i) Out crossing:

It is the mating of animals within the same breed but having no common ancestors on either side of their pedigree up to 4-6 generations. The offspring of such a cross is called as an outcross. Outcrossing is the best breeding method for animals that are below average in productivity in milk production, growth rate in beef cattle, etc. Sometimes only one outcross helps to overcome inbreeding depression.

(ii) Cross-breeding:

In cross-breeding superior males of one breed are mated with superior females of another breed. Many new animal breeds have been developed by this strategy. It gives better breeds. Cows of an inferior breed may be mated to bulls of a superior breed to get better progeny. Hisardale is a new breed of sheep developed in Punjab by crossing Bikaneri ewes and Marino rams.

(iii) Interspecific Hybridisation:

In this approach, male and female animals of two different species are mated. The progeny obtained from such a mating are usually different from both the parental species.

But in some cases, the progeny may combine desirable characters of both the parents. Mule is produced from a cross between female horse (mare) and male donkey. Mules are harder than their parents and are well suited for hard work in mountainous regions.

Controlled Breeding Experiments

These are carried out using artificial insemination and Multiple Ovulation Embryo Transfer Technology (MOET).

(i) Artificial Insemination (AI):

The semen of superior male is collected and injected into the reproductive tract of the selected female by the breeder. The semen can be used immediately or can be frozen for later use. When a bull inseminates a cow naturally approximately 5 to 10 billion sperms are deposited in the vagina. However, when semen is deposited artificially, considerably fewer sperms are required to achieve conception. Therefore, artificial insemination is very economical. The spread of certain diseases can be controlled by this method.

(ii) Multiple Ovulation Embryo Transfer Technology (MOET):

In this method, hormones (with FSH-like activity) is given to the cow for inducing follicular maturation and super ovulation instead of one egg, which they usually give per cycle, they produce 6-8 eggs. The cow is either mated with a best bull or artificially inseminated.

The embryos at 8-32 cell stage are recovered and transferred to surrogate mothers. The genetic mother is available for another super ovulation. MOET has been done in cattle, sheep, rabbits, buffaloes, mares, etc. High milk giving breeds of females and high quality (lean meat with less lipid) meat- giving bulls have been bred successfully to obtain better breed in a short time.

Selective Breeding

Natural selection and selective breeding can both produce changes in animals and plants. The difference between the two is that natural selection occurs in nature, but selective breeding only occurs when humans intervene.

Selective breeding is a process where we choose the characteristics we want in an animal. We then breed together a male and female that show some of those characteristics. From the offspring produced we select those that show the characteristic the most, and breed them together.

This process is repeated over many generations, each time selecting and breeding together those animals that have the characteristics we are looking for. Over a large number of generations, this can produce some surprising results.

The procedure involves identifying certain desirable features and finding two members of a species that exhibit the particular feature. A series of matings or breedings is then performed between the individuals with favored features to produce offspring that exhibit the feature and that can be used for future matings. The desirable *phenotypic traits* are passed from parents to offspring via their *genes*.

The term 'artificial selection' was first coined by *Charles Darwin* in his book *On the Origin of Species* to describe how humans had mirrored the process of natural selection through selective breeding.

Plants and Livestock

Almost all of the food consumed by modern humans has been selectively bred over thousands of years. Around 10,000 years ago when humans began living in permanent or semi-permanent settlements, they started to cultivate their own crops and herd flocks of livestock for the first time. Humans originally selected fruits and vegetables for qualities such as large size and sweetness unconsciously; the seeds of plants with the desirable qualities would have been given the chance to *germinate* through human consumption and cultivated within their latrines (toilets). Over time, other favorable qualities, such as oil content, seedlessness and fleshy texture, were all altered, leaving many human-cultivated fruit and vegetables unrecognizable against their wild counterparts. The same process occurred with domesticated animals such as sheep (bred for thicker wool), chickens (considerably larger than their wild ancestors), and cattle (bred for more muscle mass or increased milk yield).

One of the oldest and most widely documented examples of selective breeding for food is the selection of tall growing (for easier harvesting), disease resistant wheat, which yields large amounts of grain. Historically, smaller crops were removed from fields, allowing bees and other pollinators to pollinate only the crops with the most human-favored characteristics. Today, the breeding of wheat is a more scientific process; individuals with specific genes are identified and bred to create plants that have improved nutritional content and more intense flavors, and require less fertilizer or pesticide applications.

Selective breeding of domesticated animals has also resulted in the generation of diverse breeds of offspring:

- Examples of selective breeding of domesticated animals can be seen in horse and cows.

Example 1: Horse Breeding:

Horses have been selectively bred across many generations to produce variation according to a targeted function:

- Race horses have been bred for speed and hence are typically leaner, lighter, taller and quicker.
- Draft horses have been bred for power and endurance and hence are sturdier and stockier.

Example 2: Cow Breeding:

Cows have been selectively bred across many generations to produce offspring with improved milk production.

Farmers have also targeted the breeding a cows with a mutation resulting in increased muscle mass:

- The resulting stock of cattle (termed Belgian Blue) have excessive bulk and produce more edible lean meat.

Inbreeding

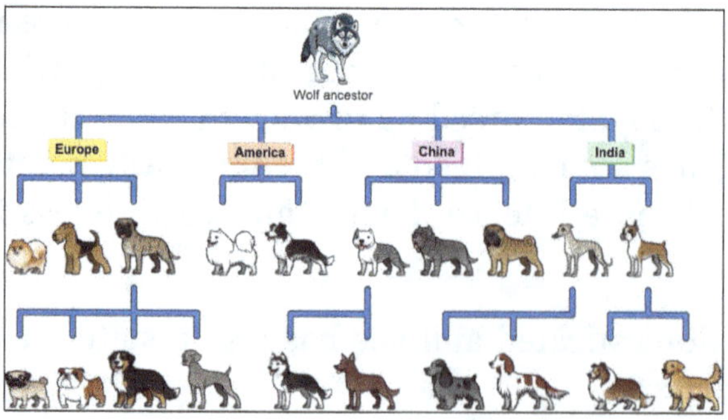

Various mating schemes of animals are classified under two broad categories — inbreeding and outbreeding. Classification depends on the closeness of the biological relationship between mates. Within each category, a wide variation in intensity of this relationship exists. A very fine line separates the two categories. Mating closely related animals (for example, parent and offspring, full brother and sister or half brother and sister) is inbreeding. With less closely related animals (first cousins, second cousins), people disagree about where to draw the line between inbreeding and outbreeding.

Technically, inbreeding is defined as the mating of animals more closely related than the average relationship within the breed or population concerned. Matings between animals less closely related than this, then, would constitute outbreeding. These two systems of mating, with varying intensities in each, are described in Table. Matings indicated within the inbreeding category are self-explanatory; those within the outbreeding category are defined in the glossary.

Degrees of inbreeding and outbreeding arranged according to biological relationship between indicated mates. (In reading from top to bottom, biological relationship between mates steadily decreases.)

Inbreeding or outbreeding	Coefficient of relationship between mates	Biological relationship between mates
Inbreeding	50 percent	Parent × offspring; full sibs
Inbreeding	25 percent	Half-sibs; double first cousins; aunt × nephew; uncle × niece
Inbreeding	12-1/2 percent	First cousins
Inbreeding	6-1/4 percent	Second cousins
Inbreeding	?	Linebreeding[1]
Inbreeding/ outbreeding	0 percent	Random mating within breed or population[2]
Outbreeding	0 percent	Outcrossing
Outbreeding	0 percent	Breed crossing
Outbreeding	0 percent	Species crossing
Outbreeding	0 percent	Genus crossing

[1]In a linebreeding program, the coefficient of relationship between mates is usually low; however, it can be quite variable.

[2]Random mating within a breed or population means that mates are chosen by chance. It should be understood that under this circumstance it is possible that either inbreeding or outbreeding could occur.

Biological Relationships between Animals

Individuals are considered to be biologically related when they have one or more common ancestors. For practical purposes, if two individuals have no common ancestor within the last five or six generations, they are considered unrelated.

Biological relationship is important in animal breeding because the closer the relationship, the higher the percentage of like genes the two individuals carry. Closeness of relationship is determined by three factors:

- How far back in the two animals' pedigrees the common ancestor appears.
- How many common ancestors they have.
- How frequently the common ancestors appear. It is also influenced by any inbreeding of the common ancestor or ancestors.

Measurement of Degree of Biological Relationship

The coefficient of relationship is a single numerical value that considers all the above-mentioned factors. It is a measure of the degree to which the genotypes (genetic constitutions) of the two animals are similar. It is estimated by the expression:

$$R_{BC} = \text{sigma}[(1/2)^{n+n'}(1 + F_A)] \div \text{Square Root of } (1 + F_B)(1 + F_C)$$

Equation 1 where:

R_{BC} = the coefficient of relationship between animals B and C which we want to measure.

sigma = the Greek symbol meaning "add."

(1/2) = the fraction of an individual's genetic material that is transmitted to its progeny. It is used in the calculation of the coefficient of relationshop because it represents the probability that, in any one generation, an identical gene from a given pair of genes is transitted to each of two particular progeny. It is also the probability that an unlike gene from a given pair of genes is transmitted to the two progeny.

n = the number of gnerations between animal B and the common ancestor.

n' = the number of generations between animal C and the common ancestor.

F_A, F_B, F_C = inbreeding coefficients of the common ancestor and of animals B and C, respectively.

If none of the animals is inbred, the coefficient of relationship is estimated as:

$$R_{BC} = \text{sigma } [(1/2)^{n+n'}]$$

Equation 2:

The use of this expression can be demonstrated with the full-sib sample pedigree and arrow diagram. In this example, assume neither sire nor dam is inbred. The arrow diagram on the right shows paths of gene flow from each of the common ancestors (D and E) to the animals whose coefficient of relationship we are measuring (B and C).

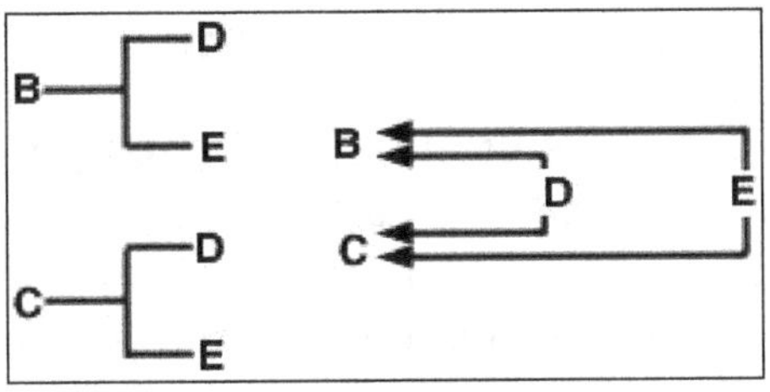

The problem now is to trace all possible paths from animal B to animal C which pass through a common ancestor. In this case, there are two such paths.

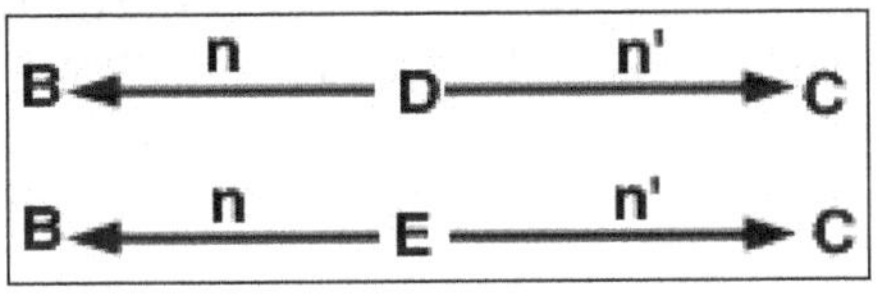

Equation 3

Since we have assumed no inbreeding in this example, the coefficient of relationship between animals B and C is estimated as:

RBC = sigma[(1/2)n + n'] = (1/2)1 + 1 + (1/2)1 + 1 = (1/2)2 + (1/2)2 = 0.50

Usefulness of Coefficient of Relationship Information

A livestock producer would find coefficient of relationship information valuable in a number of situations. He may, for example, want to sell an animal related to one that previously sold for a high price. The higher the coefficient of relationship between the two, the better its use as a sales point. Or, he may want to purchase one of two related bulls and one may cost more than he wishes to pay. If the coefficient of relationship between the two bulls is high, he might be as well off with the lower priced bull as he would with the more expensive one.

A practical use of the coefficient of relationship is estimating the performance value of an untested animal. To estimate the value, we must know the performance value of a related animal, the coefficient of relationship between the tested and untested animals, and the average performance value of the breed, herd, or group to which the tested and untested animals belong.

As an example, consider a herd with a feedlot average daily gain (ADG) of 2.25 pounds per day. Assume further that a sire from this herd had a 3.50 pounds per day feedlot ADG, while a younger half brother has not, as yet, been evaluated for feedlot ADG. Assuming no inbreeding, the coefficient of relationship between half brothers is 0.25. The best estimate of the untested animal's feedlot ADG is that it will deviate from the herd average 25 percent as far as does the performance value of the tested half brother. Using these figures, the most probable feedlot ADG value of the untested animal is 2.25 + (0.25) × (3.50 - 2.25) or 2.56 pounds per day.

Measurement of the Degree of Inbreeding

When we calculate an inbreeding coefficient, we are attempting to measure the probable percentage reduction in the frequency of pairing of dissimilar genes (reduction in heterozygosity). This reduction is relative to a base population. The base population usually is the breed concerned at a date to which the pedigrees are traced. Animals in this base population are assumed to be non-inbred. This does not mean these base population animals had dissimilar genes in each pair. There is no way for us to know how many of their gene pairs consisted of similar or dissimilar genes. The inbreeding coefficient that is calculated is simply relative to that base and reflects the probable percentage reduction in however many dissimilar gene pairs the average base population animals had.

The general expression for determining the inbreeding coefficient is:

$$F_X = \text{sigma}[(1/2)^{n+n'+1}(1 + F_A)]$$

Equation 4 where:

F_X = the inbreeding coefficient of animal X.

sigma = the Greek symbol meaning "add".

(1/2) = the fraction of an individual's genetic material that is transmitted to its progeny. It is used in the calculation of the coefficient of relationshop because it represents the probability that, in any *one* generation, an identical gene from a given pair of genes is transitted to each of two particular progeny. It is also the probability that an *unlike* gene from a given pair of genes is transmitted to the two progeny.

n = the number of generations between animal B and the common ancestor.

n' = the number of generations between animal C and the common ancestor.

+1 = is added to n and n' to account for the additional generation between animal × and its parents.

F_A = the inbreeding coefficient of the common ancestor.

If neither parent is inbred, but if they are related, the inbreeding coefficient of their progeny is half their coefficient of relationship: $1/2\ R_{BC}$ This can be demonstrated using a full-sib mating to make comparison easy with the full-sib coefficient of relationship calculated previously. In this case, the pedigree for animal X and the arrow diagram will be as follows:

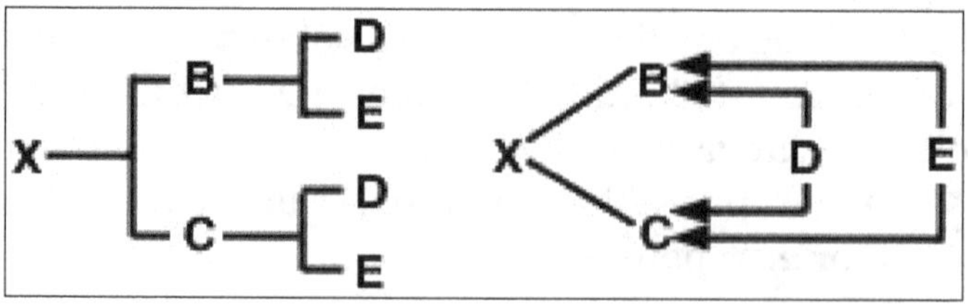

The problem now is to trace all possible paths from the sire (B) to the dam (C) through each common ancestor. As with the coefficient of relationship problem, there are two such paths:

$$X \leftarrow B \xleftarrow{n} D \xrightarrow{n'} C \rightarrow X$$
$$X \leftarrow B \xleftarrow{n} E \xrightarrow{n'} C \rightarrow X$$

Since we have assumed neither parent is inbred, the inbreeding coefficient of animal × is estimated as:

$$F_X = \text{sigma}[(1/2)^{n+n'+1}] = (1/2)^{1+1+1} + (1/2)^{1+1+1} = (1/2)^3 + (1/2)^3 = 0.25$$

Equation 5:

This is one-half the coefficient of relationship between full-sibs when there is no inbreeding.

Genetic Consequences of Inbreeding

The basic genetic consequence of inbreeding is to promote what is technically known as homozygosity. This means there is an increase in the frequency of pairing of similar genes. Accompanying this increase, there must be a decrease in the frequency of pairing of dissimilar genes. This is called a decrease in heterozygosity. These simultaneous events are the underlying reasons for the general effects on performance we observe with inbreeding.

Reasons for Inbreeding

Development of highly productive inbred lines of domestic livestock is possible. To date, however, such attempts have met with little apparent success. Although occasional high performance animals are produced, inbreeding generally results in an overall reduction in performance. This reduction is manifested in many ways. The most obvious effects of inbreeding are poorer reproductive efficiency including higher mortality rates, lower growth rates and a higher frequency of hereditary abnormalities. This has been shown by numerous studies with cattle, horses, sheep, swine and laboratory animals.

The extent of this decrease in performance, in general, is in proportion to the degree of inbreeding. The greater the degree of inbreeding, the greater the reduction in performance. The actual performance reduction is not the same in all species or in all traits. Some characteristics (like meat quality) are hardly influenced by inbreeding; others (like reproductive efficiency) are greatly influenced by inbreeding. We cannot, then, make a generalized statement about the amount of reduction in "performance" that would result from a specific amount of inbreeding and expect it to be applicable in a broad variety of situations.

It is possible, however, to predict the extent of the effect of inbreeding on specific traits. Such predictions are based on results actually obtained under experimental conditions in which various levels of inbreeding had been attained. In research with swine conducted at the Midwest Regional Swine Breeding Laboratory, Dickerson and others (1954) point out that for each 10 percent increase in inbreeding (of the pigs in the litter), there is a decrease of 0.20, 0.35, 0.38, and 0.44 pigs per litter at birth, 21 days, 56 days, and 154 days, respectively. We can use such figures to provide estimates of expected decreases in litter size (at comparable litter ages) in other herds of swine.

Most inbreeding studies suggest each successive unit increase in inbreeding results in a proportional decrease in performance. Estimates of average increases in percentage inbreeding within a closed herd can be made with the expression:

> Avg. increase in inbreeding = (Number males + Number females) ÷ (8)(Number males)(Number females).

In a closed herd of cattle in which 100 females and four males were used in each generation, for example, the average per generation increase in inbreeding would be (4 + 100) ÷ (8)(4)(100) = 0.0325. On a per year basis, assuming a generation interval of five years, this would amount to an average yearly increase in inbreeding of 0.0065 or 0.65 percent.

Despite the generally poor results obtained with inbreeding, it is a very useful tool in animal breeding. Inbreeding is essential to the development of prepotent animals — animals that uniformly "stamp" their characteristics on their progeny. Because inbreeding causes an increase in the proportion of like genes (good or bad, recessive or dominant), the inbred animal's reproductive cells will be more uniform in their genetic makeup. When this uniformity involves a relatively large number of dominant genes, the progeny of that individual will uniformly display the dominant characteristics of that parent.

Inbreeding may also be used to uncover genes that produce abnormalities or death — genes that, in outbred herds, are generally present in low frequencies. These harmful genes are almost always recessive in their genetic nature and their effects are hidden or masked by their dominant counterparts (alleles). Except for sex-linked traits, recessive genes are not expressed if carried singly. For their effects to be manifested, they must be present in duplicate. The likelihood they will be present in duplicate increases with inbreeding, because inbreeding increases the proportion of like genes (both good and bad) in the inbred population. With the effects of these genes uncovered, the breeder can eliminate them from his herd. He would cull progeny that showed the undesirable effect of these recessive genes and would also cull the parents that are carriers of the undesirable genes. In addition, two-thirds of the "normal" progeny of these carrier parents are themselves expected to be carriers of these same undesirable genes. In the absence of breeding tests to sort out the carriers from the non-carriers, it would also be necessary to cull all "normal" progeny of the carrier parents.

Breeders can use an inbreeding test to identify carriers of harmful autosomal recessive genes (like those responsible for snorter dwarfism in cattle, hyperostosis in swine, or cryptorchidism in sheep). An inbreeding test checks for only recessive genes that the tested animal (usually the male) carries. The following example and the numbers given are pertinent to cattle, where one offspring per gestation is usual. If we want to be sure at the 0.01 level of probability that a bull is free of harmful autosomal recessive genes, we would have to mate him to at least 35 of his daughters. He is mated to 35+ of his daughters because only half of them are expected to carry any harmful recessive gene their sire is carrying. We need the production of 35 normal calves *without a single abnormal calf* to show the bull free (at the 0.01 probability level) of any harmful autosomal recessive gene.

Tests in sheep and swine would require matings to fewer daughters. Actual numbers would depend on the average number of progeny produced per gestation in each species. Using the 0.01 probability level and assuming the average litter size is 1.5 in sheep and 8.0 in swine, the number of sire-daughter matings needed would be about 24+ for sheep and 5+ for swine.

Another important use of inbreeding is in the development of distinct families or inbred lines. Beginning with an initially diverse genetic population, inbreeding results in the formation of various lines, each differing genetically from the other. Continued inbreeding within these lines tends to change the frequency of some of the genes found in the initial population. For example, if a particular gene is present in only 1 percent of the animals in the initial population, inbreeding and the development of distinct lines could result in this gene being present in all or nearly all animals in some lines and in none or only a few of the animals in other lines. Inbred lines are used in a number of ways but are probably most notably used in the development of hybrid chickens or hybrid seed corn.

A generally mild form of inbreeding (linebreeding) is being used successfully by some seed stock and commercial producers. Its objective is to maintain a high degree of relationship between the animals in the herd and some outstanding ancestor or ancestors. With inbreeding in general, there is no attempt to increase the relationship between the offspring and any particular ancestor. In a linebreeding program there is a deliberate attempt to maintain or increase the relationship between the offspring and a specific admired ancestor (or ancestors). This feature distinguishes linebreeding as a special form of inbreeding.

The inbreeding coefficient of the offspring produced in a linebreeding program is generally low, but is dependent on the kind of breeding program followed. Figure illustrates two linebreeding programs. The breeding program outlined in part A shows a direct line of relationship between offspring "x" and desired ancestor "5." Only a mild level of inbreeding in animal "x" (F_x = 0.03125) is reached. The coefficient of relationship obtained between animal "x" and animal "5" is 0.2462. Part B is a linebreeding program based on continuous sire-daughter matings which, after only two generations, produces a level of inbreeding in animal "x" of 0.375 and a coefficient of relationship between animals "x" and "S" of 0.78.

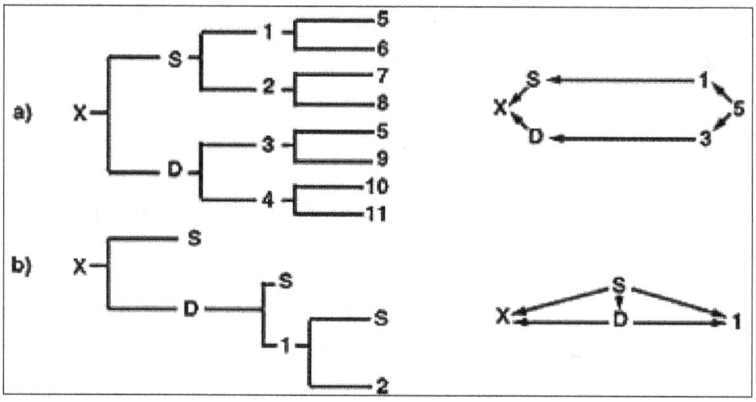

Pedigrees and arrow diagrams for two line breeding programs.

The sire-daughter program in part B effectively concentrates genes from animal "S," but because of the rapid increase in inbreeding of the progeny produced in this program, the breeder runs the risk of greatly reduced performance and a high probability of genetic defects. Linebreeding programs are best used in purebred and high performing herds when a truly genetically superior individual in that herd has been identified and evaluated by progeny testing. Concentrating that individual's genes would then be best accomplished by mating him to unrelated females to reduce the risk of harmful effects associated with such intense inbreeding.

Outbreeding

Out-crossing or out-breeding means that the crossing between different breeds. This is the practice of introducing unrelated genetic material into a breeding line. It increases genetic diversity, thus reducing the probability of an individual being subject to disease or genetic abnormalities.

Outcrossing is now the norm of most purposeful animal breeding, contrary to what is commonly believed. The outcrossing breeder intends to remove the traits by using "new blood". With dominant traits, one can still see the expression of the traits and can remove those traits whether one outcrosses, line breeds or inbreds. With recessive traits, outcrossing allows for the recessive traits to migrate across a population. The outcrossing breeder then may have individuals that have many deleterious genes that may be expressed by subsequent inbreeding. There is now a gamut of deleterious genes within each individual in many dog breeds.

Increasing the variation of genes or alleles within the gene pool may protect against extinction by stressors from the environment. For example, in this context, a recent veterinary medicine study tried to determine the genetic diversity within cat breeds.

Outcrossing is believed to be the “norm” in the wild. Outcrossing in plants is usually enforced by self-incompatibility.

Breeders inbreed within their genetic pool, attempting to maintain desirable traits and to cull those traits that are undesirable. When undesirable traits begin to appear, mates are selected to determine if a trait is recessive or dominant. Removal of the trait is accomplished by breeding two individuals known not to carry it.

Gregor Mendel used outcrossing in his experiments with flowers. He then used the resulting offspring to chart inheritance patterns, using the crossing of siblings, and backcrossing to parents to determine how inheritance functioned.

Charles Darwin, in his book The Effects of Cross and Self-Fertilization in the Vegetable Kingdom, came to clear and definite conclusions concerning the adaptive benefit of outcrossing. For example, he stated (on page 462) that “the offspring from the union of two distinct individuals, especially if their progenitors have been subjected to very different conditions, have an immense advantage in height, weight, constitutional vigor and fertility over the self-fertilizing offspring from either one of the same parents”. He thought that this observation was amply sufficient to account for outcrossing sexual reproduction. The disadvantages of self-fertilized offspring (inbreeding depression) are now thought to be largely due to the homozygous expression of deleterious recessive mutations; and the fitness advantages of outcrossed offspring are thought to be largely due to the heterozygous masking of such deleterious mutations.

Outbreeding Depression

In biology, outbreeding depression is when progeny resulting from crosses between genetically distant individuals (outcrossing) exhibit lower fitness in the parental environment than either of their parents or than progeny from crosses between individuals that are more closely related. The concept is opposed to inbreeding depression, although the two effects can occur simultaneously. Outbreeding depression manifests most significantly in two ways:

- Intermediate genotypes are not adapted to either parental habitat. For example, selection in one population might favor a large body size, whereas in another population small body size might be more advantageous, while individuals with intermediate body sizes are comparatively disadvantaged in both populations. As another example, in the Tatra Mountains, the introduction of ibex from the Middle East resulted in hybrids which produced calves at the coldest time of the year.

- Breakdown of biochemical or physiological compatibility. Within isolated breeding populations, alleles are selected in the context of the local genetic background. Because the same alleles may have rather different effects in different genetic backgrounds, there is the potential evolution of different locally adapted gene complexes. Outcrossing between individuals with differently adapted gene complexes can result in disruption of this selective advantage, resulting in a loss of fitness.

Mechanism and Impact

The different mechanisms of outbreeding depression can operate at the same time. However, determining which mechanism is more important in a particular population is very difficult. Generally the first mechanism will be more prevalent in the first generation (F1) after the initial outcrossing when most individuals are made up of the intermediate phenotype. An extreme case of this type of outbreeding depression is the sterility and other fitness-reducing effects often seen in interspecific hybrids (such as mules), which involves not only different alleles of the same gene but even different orthologous genes.

The second mechanism may not appear until two or more generations later (F2 or greater), when recombination has undermined vitality positive epistasis. Hybrid vigor in the first generation can, in some circumstances, be strong enough to mask the effects of outbreeding depression. An example of this is that plant breeders will make F1 hybrids from purebred strains, which will improve the uniformity and vigor of the offspring, however the F1 generation are not used for further breeding because of unpredictable phenotypes in their offspring. Unless there is strong selective pressure, outbreeding depression can increase in further generations as co-adapted gene complexes are broken apart without the forging of new co-adapted gene complexes to take their place.

If the outcrossing is limited and populations are large enough, selective pressure acting on each generation can restore fitness. Unless the F1 hybrid generation is sterile or very low fitness, selection will act in each generation using the increased diversity to adapt to the environment. This can lead to recovery in fitness to baseline, and sometimes even greater fitness than original parental types in that environment. However, as the hybrid population will likely to go through a decline in fitness for a few generations, they will need to persist long enough to allow selection to act before they can rebound.

Crossbreeding

Crossbreeding is the deliberate mating of animals from different breeds or strains designed to take advantage of Heterosis (Hybrid Vigour) for characteristics like production, fertility and longevity.

Heterosis (Hybrid Vigour)

Heterosis/Hybrid Vigour is the tendency of a crossbred animal to have qualities superior to that of either parent but not more than the dominant breed.

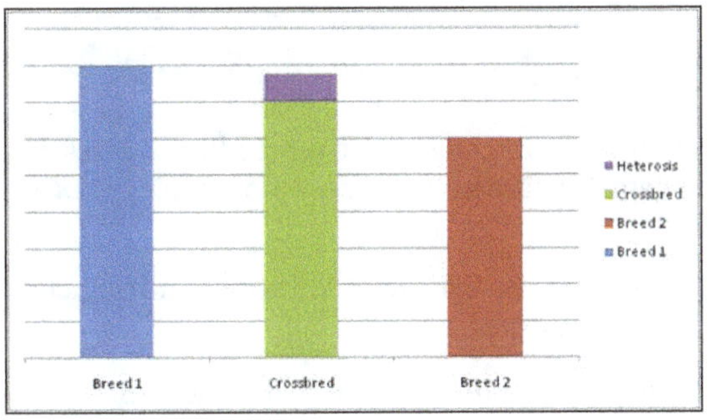

The above example of heterosis can be applied to all health or production traits if there is sufficent genetic difference between the two chosen breeds. For the above example if Breed 1 was crossed with Breed 2 the green area of the crossbred would be the logical average animal produced by the two breeds. The purple area on top of this is the added benefit or heterosis effect. The heterosis effect is variable between breeds and also the traits affected.

Trial work has shown the main benefit of heterosis is seen in fitness and health traits. 10% to 18% heterosis has been recorded in trials carried out in California and New Zealand in a wide variety of crossbred cows including Swedish/Norwegian Reds (SRB), Brown Swiss, Jerseys, and Montbelliarde. Milk production has been recorded as improving up to 6% but it is important to remember this is not the only measure of profitability.

To gain and maintain the effects of heterosis, breed and bull selection is paramount. To maximise the benefits farmers need to plan the direction they want their business to take in the long term to ensure cross breeding is the right path for them. Farmers initially need to look at the traits they wish to improve within their herd and then select bulls using all avaialble bull proofs including PLI and PIN.

2 Way Cross	67%
3 Way Cross	85%
4 Way Cross	94%

Although working a four way cross system provides maximum gain from heterosis, it is widly recomended a three way cross is more sustainable. Using a four way cross is:

- Difficult to implement.
- Choosing four suitable breeds can be a problem.
- Breed qualities are watered down with each cross.

Heterosis should not be used as the main reason to crossbreed but as an added bonus. It is important to remember that crossbreeding is about herd improvement not breed improvement.

References

- Fenster, Charles (2000). "Inbreeding and Outbreeding Depression in Natural Populations of Chamaecrista fasciculata (Fabaceae)". Conservation Biology. 14 (5): 1406–1412. doi:10.1046/j.1523-1739.2000.99234.x
- Frankham, Ballou, & Briscoe, R., J.D. & D.A. (2002). Introduction to Conservation Genetics. Cambridge. p. 382. ISBN 0521702712
- Turcek, FJ (1951). "Effect of introductions on two game populations in Czechoslovakia". Journal of Wildlife Management. 15: 113–114
- Lewontin & Birch, R.C. & L.C. (February 3, 1966). "Hybridization as a source of variation for adaptation to new environments". Evolution. 20 (3): 315–336. doi:10.2307/2406633. JSTOR 2406633. PMID 28562982
- Bernstein H, Hopf FA, Michod RE (1987). "The molecular basis of the evolution of sex". Adv. Genet. 24: 323–70. doi:10.1016/s0065-2660(08)60012-7. PMID 3324702

5

Animal Products

Any material derived from the body of an animal is termed as animal product. They are broadly classified as food and non-food products. A few of the products which are derived from animals are meat, milk, wool, fur, fiber, etc. This capter has been carefully written to provide an extensive understanding of the different aspects of these animal products.

Food

Food is any substance consumed to provide nutritional support for the body. It is usually of plant or animal origin, and contains essential nutrients, such as fats, proteins, vitamins, or minerals. The substance is ingested by an organism and assimilated by the organism's cells to provide energy, maintain life, or stimulate growth.

Historically, people secured food through two methods: hunting and gathering and agriculture. Today, the majority of the food energy required by the ever increasing population of the world is supplied by the food industry.

Food safety and food security are monitored by agencies like the International Association for Food Protection, World Resources Institute, World Food Programme, Food and Agriculture Organization, and International Food Information Council. They address issues such as sustainability, biological diversity, climate change, nutritional economics, population growth, water supply, and access to food.

The right to food is a human right derived from the International Covenant on Economic, Social and Cultural Rights (ICESCR), recognizing the "right to an adequate standard of living, including adequate food", as well as the "fundamental right to be free from hunger".

Various foods

Foods from plant sources.

Food Sources

Most food has its origin in plants. Some food is obtained directly from plants; but even animals that are used as food sources are raised by feeding them food derived from plants. Cereal grain is a staple food that provides more food energy worldwide than any other type of crop. Corn (maize), wheat, and rice – in all of their varieties – account for 87% of all grain production worldwide. Most of the grain that is produced worldwide is fed to livestock.

Some foods not from animal or plant sources include various edible fungi, especially mushrooms. Fungi and ambient bacteria are used in the preparation of fermented and pickled foods like leavened bread, alcoholic drinks, cheese, pickles, kombucha, and yogurt. Another example is blue-green algae such as Spirulina. Inorganic substances such as salt, baking soda and cream of tartar are used to preserve or chemically alter an ingredient.

Plants

Many plants and plant parts are eaten as food and around 2,000 plant species are cultivated for food. Many of these plant species have several distinct cultivars.

Seeds of plants are a good source of food for animals, including humans, because they contain the nutrients necessary for the plant's initial growth, including many healthful fats, such as Omega fats. In fact, the majority of food consumed by human beings are seed-based foods. Edible seeds include cereals (corn, wheat, rice, et cetera), legumes (beans, peas, lentils, et cetera), and nuts. Oilseeds are often pressed to produce rich oils - sunflower, flaxseed, rapeseed (including canola oil), sesame, et cetera.

Seeds are typically high in unsaturated fats and, in moderation, are considered a health food, although not all seeds are edible. Large seeds, such as those from a lemon, pose a choking hazard, while seeds from cherries and apples contain cyanide which could be poisonous only if consumed in large volumes.

Fruits are the ripened ovaries of plants, including the seeds within. Many plants and animals have coevolved such that the fruits of the former are an attractive food source to the latter, because animals that eat the fruits may excrete the seeds some distance away. Fruits, therefore, make up a significant part of the diets of most cultures. Some botanical fruits, such as tomatoes, pumpkins, and eggplants, are eaten as vegetables.

Vegetables are a second type of plant matter that is commonly eaten as food. These include root vegetables (potatoes and carrots), bulbs (onion family), leaf vegetables (spinach and lettuce), stem vegetables (bamboo shoots and asparagus), and inflorescence vegetables (globe artichokes and broccoli and other vegetables such as cabbage or cauliflower).

Animals

Animals are used as food either directly or indirectly by the products they produce. Meat is an example of a direct product taken from an animal, which comes from muscle systems or from organs.

Various raw meats.

Food products produced by animals include milk produced by mammary glands, which in many cultures is drunk or processed into dairy products (cheese, butter, etc.). In addition, birds and other animals lay eggs, which are often eaten, and bees produce honey, a reduced nectar from flowers, which is a popular sweetener in many cultures. Some cultures consume blood, sometimes in the form of blood sausage, as a thickener for sauces, or in a cured, salted form for times of food scarcity, and others use blood in stews such as jugged hare.

Some cultures and people do not consume meat or animal food products for cultural, dietary, health, ethical, or ideological reasons. Vegetarians choose to forgo food from animal sources to varying degrees. Vegans do not consume any foods that are or contain ingredients from an animal source.

Production

Most food has always been obtained through agriculture. With increasing concern over both the methods and products of modern industrial agriculture, there has been a growing trend toward sustainable agricultural practices. This approach, partly fueled by consumer demand, encourages biodiversity, local self-reliance and organic farming methods. Major influences on food production include international organizations (e.g. the World Trade Organization and Common Agricultural Policy), national government policy (or law), and war.

A tractor and chaser bin.

In popular culture, the mass production of food, specifically meats such as chicken and beef, has come under fire from various documentaries, most recently Food, Inc, documenting the mass slaughter and poor treatment of animals, often for easier revenues from large corporations. Along with a current trend towards environmentalism, people in Western culture have had an increasing trend towards the use of herbal supplements, foods for a specific group of people (such as dieters, women, or athletes), functional foods (fortified foods, such as omega-3 eggs), and a more ethnically diverse diet.

Several organisations have begun calling for a new kind of agriculture in which agroecosystems provide food but also support vital ecosystem services so that soil fertility and biodiversity are maintained rather than compromised. According to the International Water Management Institute and UNEP, well-managed agroecosystems not only provide food, fiber and animal products, they also provide services such as flood mitigation, groundwater recharge, erosion control and habitats for plants, birds fish and other animals.

Taste Perception

Animals, specifically humans, have five different types of tastes: sweet, sour, salty, bitter, and umami. As animals have evolved, the tastes that provide the most energy (sugar and fats) are the most pleasant to eat while others, such as bitter, are not enjoyable. Water, while important for survival, has no taste. Fats, on the other hand, especially saturated fats, are thicker and rich and are thus considered more enjoyable to eat.

Sweet

Generally regarded as the most pleasant taste, sweetness is almost always caused by a type of simple sugar such as glucose or fructose, or disaccharides such as sucrose, a molecule combining glucose and fructose. Complex carbohydrates are long chains and thus do not have the sweet taste. Artificial sweeteners such as sucralose are used to mimic the sugar molecule, creating the sensation of sweet, without the calories. Other types of sugar include raw sugar, which is known for its amber color, as it is unprocessed. As sugar is vital for energy and survival, the taste of sugar is pleasant.

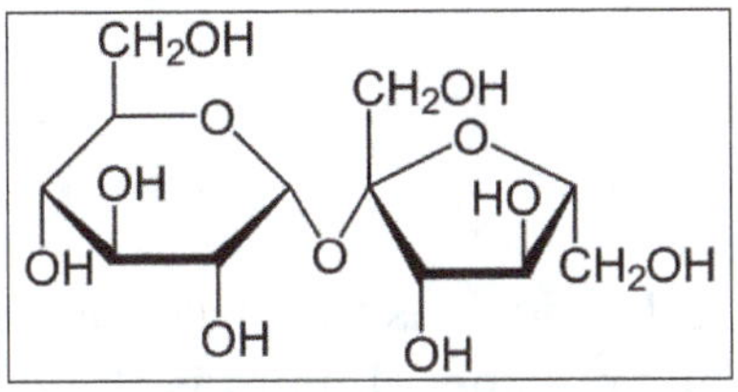

Structure of sucrose.

The stevia plant contains a compound known as steviol which, when extracted, has 300 times the sweetness of sugar while having minimal impact on blood sugar.

Sour

Sourness is caused by the taste of acids, such as vinegar in alcoholic beverages. Sour foods include citrus, specifically lemons, limes, and to a lesser degree oranges. Sour is evolutionarily significant as it is a sign for a food that may have gone rancid due to bacteria. Many foods, however, are slightly acidic, and help stimulate the taste buds and enhance flavor.

Salty

Saltiness is the taste of alkali metal ions such as sodium and potassium. It is found in almost every food in low to moderate proportions to enhance flavor, although to eat pure salt is regarded as highly unpleasant. There are many different types of salt, with each having a different degree of saltiness, including sea salt, fleur de sel, kosher salt, mined salt, and grey salt. Other than enhancing flavor, its significance is that the body needs and maintains a delicate electrolyte balance, which is the kidney's function. Salt may be iodized, meaning iodine has been added to it, a necessary nutrient that promotes thyroid function. Some canned foods, notably soups or packaged broths, tend to be high in salt as a means of preserving the food longer. Historically salt has long been used as a meat preservative as salt promotes water excretion. Similarly, dried foods also promote food safety.

Salt mounds in Bolivia.

Bitter

Bitterness is a sensation often considered unpleasant characterized by having a sharp, pungent taste. Unsweetened dark chocolate, caffeine, lemon rind, and some types of fruit are known to be bitter.

Umami

Umami, the Japanese word for delicious, is the least known in Western popular culture but has a long tradition in Asian cuisine. Umami is the taste of glutamates, especially monosodium glutamate (MSG). It is characterized as savory, meaty, and rich in flavor. Salmon and mushrooms are foods high in umami.

Cuisine

Many cultures have a recognizable cuisine, a specific set of cooking traditions using various spices or a combination of flavors unique to that culture, which evolves over time. Other differences include preferences (hot or cold, spicy, etc.) and practices, the study of which is known as gastronomy. Many cultures have diversified their foods by means of preparation, cooking methods, and manufacturing. This also includes a complex food trade which helps the cultures to economically survive by way of food, not just by consumption.

Typical Assyrian cuisine.

Some popular types of ethnic foods include Italian, French, Japanese, Chinese, American, Cajun, Thai, African, and Indian cuisine. Various cultures throughout the world study the dietary analysis of food habits. While evolutionarily speaking, as opposed to culturally, humans are omnivores, religion and social constructs such as morality, activism, or environmentalism will often affect which foods they will consume. Food is eaten and typically enjoyed through the sense of taste, the perception of flavor from eating and drinking. Certain tastes are more enjoyable than others, for evolutionary purposes.

Presentation

Aesthetically pleasing and eye-appealing food presentations can encourage people to consume foods. A common saying is that people "eat with their eyes". Food presented in a clean and appetizing way will encourage a good flavor, even if unsatisfactory.

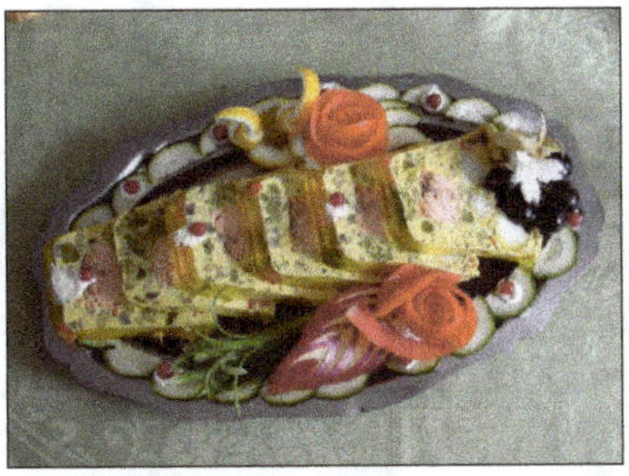

A French basil salmon terrine, with eye-appealing garnishes.

Contrast in Texture

Texture plays a crucial role in the enjoyment of eating foods. Contrasts in textures, such as something crunchy in an otherwise smooth dish, may increase the appeal of eating it. Common examples include adding granola to yogurt, adding croutons to a salad or soup, and toasting bread to enhance its crunchiness for a smooth topping, such as jam or butter.

Contrast in Taste

Another universal phenomenon regarding food is the appeal of contrast in taste and presentation. For example, such opposite flavors as sweetness and saltiness tend to go well together, as in kettle corn and nuts.

Food Preparation

While many foods can be eaten raw, many also undergo some form of preparation for reasons of safety, palatability, texture, or flavor. At the simplest level this may involve washing, cutting, trimming, or adding other foods or ingredients, such as spices. It may also involve mixing, heating or cooling, pressure cooking, fermentation, or combination with other food. In a home, most food preparation takes place in a kitchen. Some preparation is done to enhance the taste or aesthetic appeal; other preparation may help to preserve the food; others may be involved in cultural identity. A meal is made up of food which is prepared to be eaten at a specific time and place.

A refrigerator helps to keep foods fresh.

Animal Preparation

The preparation of animal-based food usually involves slaughter, evisceration, hanging, portioning, and rendering. In developed countries, this is usually done outside the home in slaughterhouses, which are used to process animals en masse for meat production. Many countries regulate their slaughterhouses by law. For example, the United States has established the Humane Slaughter Act of 1958, which requires that an animal be stunned before killing. This act, like those in many countries, exempts slaughter in accordance to religious law, such as kosher, shechita, and dhabīĥa halal. Strict interpretations of kashrut require the animal to be fully aware when its carotid artery is cut.

On the local level, a butcher may commonly break down larger animal meat into smaller manageable cuts, and pre-wrap them for commercial sale or wrap them to order in butcher paper. In addition, fish and seafood may be fabricated into smaller cuts by a fish monger. However fish butchery may be done on board a fishing vessel and quick-frozen for preservation of quality.

Cooking

The term "cooking" encompasses a vast range of methods, tools, and combinations of ingredients to improve the flavor or digestibility of food. Cooking technique, known as culinary art, generally requires the selection, measurement, and combining of ingredients in an ordered procedure in an effort to achieve the desired result. Constraints on success include the variability of ingredients, ambient conditions, tools, and the skill of the individual cook. The diversity of cooking worldwide is a reflection of the myriad nutritional, aesthetic, agricultural, economic, cultural, and religious considerations that affect it.

Cooking with a wok in China.

Cooking requires applying heat to a food which usually, though not always, chemically changes the molecules, thus changing its flavor, texture, appearance, and nutritional properties. Cooking certain proteins, such as egg whites, meats, and fish, denatures the protein, causing it to firm. There is archaeological evidence of roasted foodstuffs at *Homo erectus* campsites dating from 420,000 years ago. Boiling as a means of cooking requires a container, and has been practiced at least since the 10th millennium BC with the introduction of pottery.

Cooking Equipment

There are many different types of equipment used for cooking.

A stainless steel frying pan.

A traditional asado (barbecue).

Ovens are mostly hollow devices that get very hot (up to 500 °F) and are used for baking or roasting and offer a dry-heat cooking method. Different cuisines will use different types of ovens. For example, Indian culture uses a Tandoor oven, which is a cylindrical clay oven which operates at a single high temperature. Western kitchens use variable temperature convection ovens, conventional ovens, toaster ovens, or non-radiant heat ovens like the microwave oven. Classic Italian cuisine includes the use of a brick oven containing burning wood. Ovens may be wood-fired, coal-fired, gas, electric, or oil-fired.

Various types of cook-tops are used as well. They carry the same variations of fuel types as the ovens mentioned above. Cook-tops are used to heat vessels placed on top of the heat source, such as a sauté pan, sauce pot, frying pan, or pressure cooker. These pieces of equipment can use either a moist or dry cooking method and include methods such as steaming, simmering, boiling, and poaching for moist methods, while the dry methods include sautéing, pan frying, and deep-frying.

In addition, many cultures use grills for cooking. A grill operates with a radiant heat source from below, usually covered with a metal grid and sometimes a cover. An open pit barbecue in the American south is one example along with the American style outdoor grill fueled by wood, liquid propane, or charcoal along with soaked wood chips for smoking. A Mexican style of barbecue is called barbacoa, which involves the cooking of meats such as whole sheep over an open fire. In Argentina, an asado (Spanish for "grilled") is prepared on a grill held over an open pit or fire made upon the ground, on which a whole animal or smaller cuts are grilled.

Raw Food Preparation

Certain cultures highlight animal and vegetable foods in their raw state. Salads consisting of raw vegetables or fruits are common in many cuisines. Sashimi in Japanese cuisine consists of raw sliced fish or other meat, and sushi often incorporates raw fish or seafood. Steak tartare and salmon tartare are dishes made from diced or ground raw beef or salmon, mixed with various ingredients and served with baguettes, brioche, or frites. In Italy, carpaccio is a dish of very thinly sliced raw beef, drizzled with a vinaigrette made with olive oil. The health food movement known as raw foodism promotes a mostly vegan diet of raw fruits, vegetables, and grains prepared in various ways, including juicing, food dehydration, sprouting, and other methods of preparation that do not heat the food above 118 °F (47.8 °C). An example of a raw meat dish is ceviche, a Latin American dish made with raw meat that is "cooked" from the highly acidic citric juice from lemons and limes along with other aromatics such as garlic.

Many types of fish ready to be eaten, including salmon and tuna.

Restaurants

Restaurants employ trained chefs who prepare food, and trained waitstaff to serve the customers. The term restaurant is credited to the French from the 19th century, as it relates to the restorative nature of the bouillons that were once served in them. However, the concept pre-dates the naming of these establishments, as evidence suggests commercial food preparation may have existed during the age of the city of Pompeii, and urban sales of prepared foods may have existed in China during the Song dynasty. The coffee shops or cafés of 17th century Europe may also be considered an early version of the restaurant. In 2005, the population of the United States spent $496 billion for out-of-home dining. Expenditures by type of out-of-home dining were as follows: 40% in full-service restaurants, 37.2% in limited service restaurants (fast food), 6.6%

in schools or colleges, 5.4% in bars and vending machines, 4.7% in hotels and motels, 4.0% in recreational places, and 2.2% in others, which includes military bases.

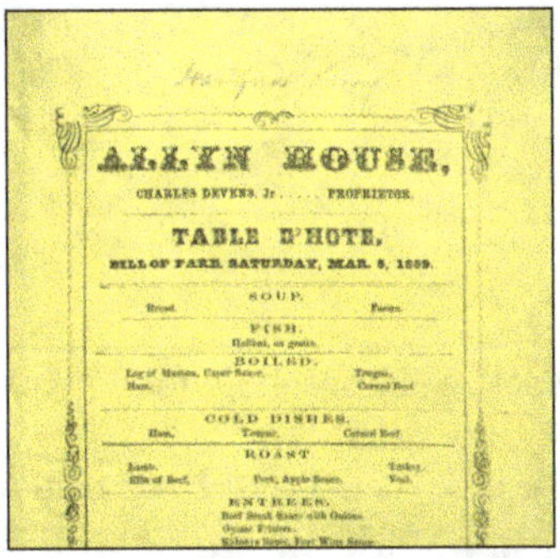

ALLYN HOUSE,

CHARLES DEVENS, Jr . . . . . PROPRIETOR.

TABLE D'HOTE,

BILL OF FARE, SATURDAY, MAR. 5, 1859.

SOUP.

FISH.

BOILED.

COLD DISHES.

ROAST

ENTREES.

The Allyn House restaurant menu (March 5, 1859).

A McDonald's restaurant in Riyadh, Saudi Arabia.

Tom's Restaurant, a restaurant in New York City.

Food Manufacturing

Packaged foods are manufactured outside the home for purchase. This can be as simple as a butcher preparing meat, or as complex as a modern international food industry. Early food processing techniques were limited by available food preservation, packaging, and transportation. This mainly involved salting, curing, curdling, drying, pickling, fermenting, and smoking. Food manufacturing arose during the industrial revolution in the 19th century. This development took advantage of new mass markets and emerging technology, such as milling, preservation, packaging and labeling, and transportation. It brought the advantages of pre-prepared time-saving food to the bulk of ordinary people who did not employ domestic servants.

Packaged household food items.

At the start of the 21st century, a two-tier structure has arisen, with a few international food processing giants controlling a wide range of well-known food brands. There also exists a wide array of small local or national food processing companies. Advanced technologies have also come to change food manufacture. Computer-based control systems, sophisticated processing and packaging methods, and logistics and distribution advances can enhance product quality, improve food safety, and reduce costs.

Commercial Trade

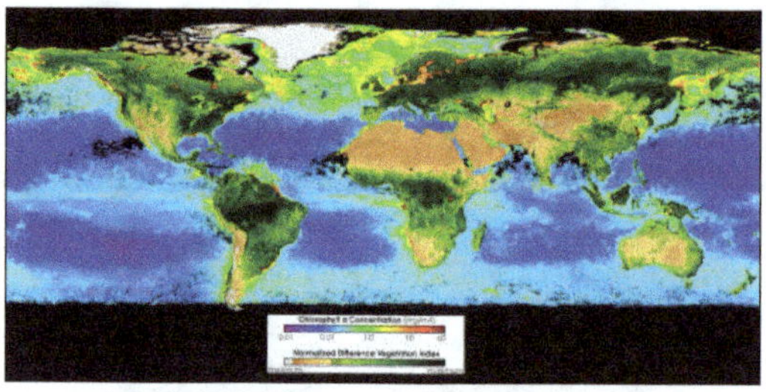

SeaWiFS image for the global biosphere.

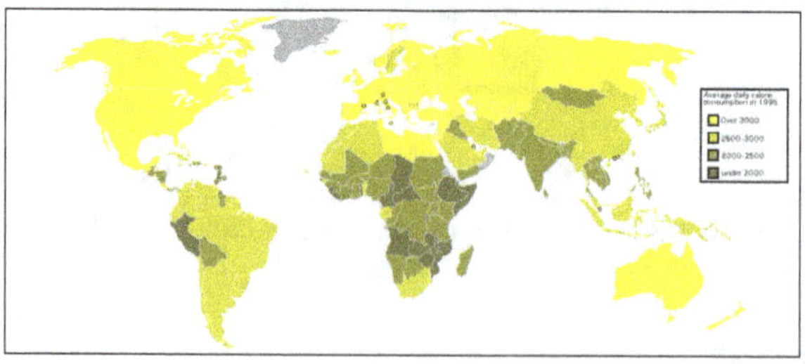

Global average daily calorie consumption in 1995.

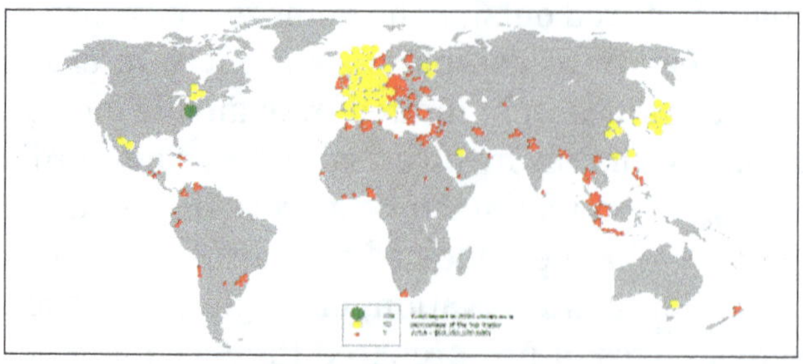

Food imports in 2005.

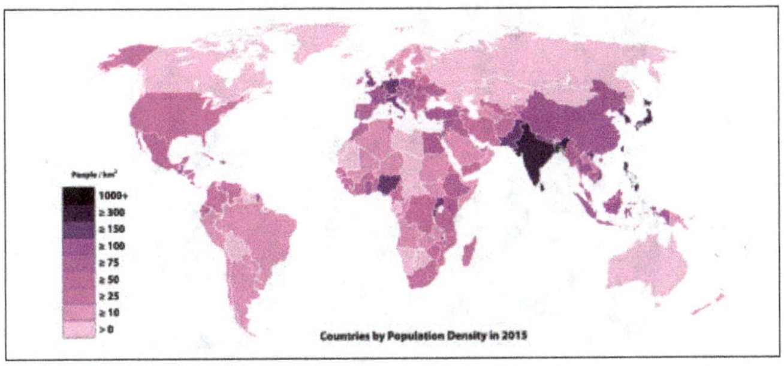

Population density of world regions.

International Food Imports and Exports

The World Bank reported that the European Union was the top food importer in 2005, followed at a distance by the USA and Japan. Britain's need for food was especially well illustrated in World War II. Despite the implementation of food rationing, Britain remained dependent on food imports and the result was a long term engagement in the Battle of the Atlantic.

Food is traded and marketed on a global basis. The variety and availability of food is no longer restricted by the diversity of locally grown food or the limitations of the local growing season. Between 1961 and 1999, there was a 400% increase in worldwide food exports. Some countries are now economically dependent on food exports, which in some cases account for over 80% of all exports.

In 1994, over 100 countries became signatories to the Uruguay Round of the General Agreement on Tariffs and Trade in a dramatic increase in trade liberalization. This included an agreement to reduce subsidies paid to farmers, underpinned by the WTO enforcement of agricultural subsidy, tariffs, import quotas, and settlement of trade disputes that cannot be bilaterally resolved. Where trade barriers are raised on the disputed grounds of public health and safety, the WTO refer the dispute to the Codex Alimentarius Commission, which was founded in 1962 by the United Nations Food and Agriculture Organization and the World Health Organization. Trade liberalization has greatly affected world food trade.

Marketing and Retailing

Food marketing brings together the producer and the consumer. It is the chain of activities that brings food from "farm gate to plate". The marketing of even a single food product can be a complicated process involving many producers and companies. For example, fifty-six companies are involved in making one can of chicken noodle soup. These businesses include not only chicken and vegetable processors but also the companies that transport the ingredients and those who print labels and manufacture cans. The food marketing system is the largest direct and indirect non-government employer in the United States.

Packaged food aisles of supermarket in Portland, Oregon, United States of America.

In the pre-modern era, the sale of surplus food took place once a week when farmers took their wares on market day into the local village marketplace. Here food was sold to grocers for sale in their local shops for purchase by local consumers. With the onset of industrialization and the development of the food processing industry, a wider range of food could be sold and distributed in distant locations. Typically early grocery shops would be counter-based shops, in which purchasers told the shop-keeper what they wanted, so that the shop-keeper could get it for them.

In the 20th century, supermarkets were born. Supermarkets brought with them a self service approach to shopping using shopping carts, and were able to offer quality food at lower cost through economies of scale and reduced staffing costs. In the latter part of the 20th century, this has been further revolutionized by the development of vast warehouse-sized, out-of-town supermarkets, selling a wide range of food from around the world.

Unlike food processors, food retailing is a two-tier market in which a small number of very large companies control a large proportion of supermarkets. The supermarket giants wield great purchasing power over farmers and processors, and strong influence over consumers. Nevertheless, less than 10% of consumer spending on food goes to farmers, with larger percentages going to advertising, transportation, and intermediate corporations.

Prices

Some essential food products including bread, rice and pasta.

It was reported on March 24, 2008, that consumers worldwide faced rising food prices. Reasons for this development include changes in the weather and dramatic changes in the global economy, including higher oil prices, lower food reserves, and growing consumer demand in China and India. In the long term, prices are expected to stabilize. Farmers will grow more grain for both fuel and food and eventually bring prices down. Already this is happening with wheat, with more crops to be planted in the United States, Canada, and Europe in 2009. However, the Food and Agriculture Organization projects that consumers still have to deal with more expensive food until at least 2018.

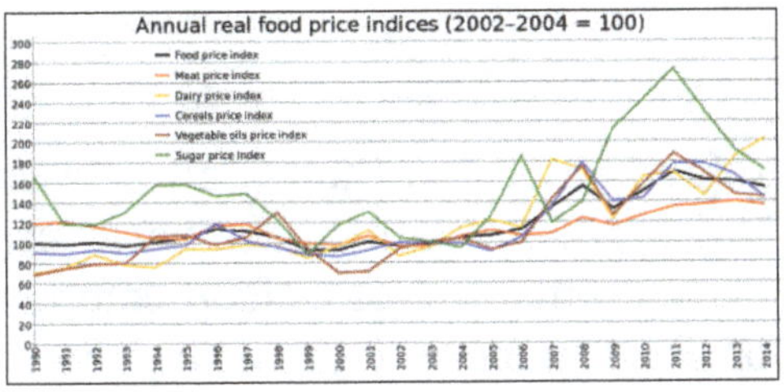

Food, meat, dairy, cereals, vegetable oil, and sugar price indices, deflated using the World Bank Manufactures Unit Value Index (MUV).

It is rare for the spikes to hit all major foods in most countries at once. Food prices rose 4% in the United States in 2007, the highest increase since 1990, and are expected to climb as much again in 2008. As of December 2007, 37 countries faced food crises, and 20 had imposed some sort of food-price controls. In China, the price of pork jumped 58% in 2007. In the 1980s and 1990s, farm subsidies and support programs allowed major grain exporting countries to hold large surpluses, which could be tapped during food shortages to keep prices down. However, new trade policies have made agricultural production much more responsive to market demands, putting global food reserves at their lowest since 1983.

Food prices are rising, wealthier Asian consumers are westernizing their diets, and farmers and nations of the third world are struggling to keep up the pace. The past five years have seen rapid growth in the contribution of Asian nations to the global fluid and powdered milk manufacturing industry, which in 2008 accounted for more than 30% of production, while China alone accounts for more than 10% of both production and consumption in the global fruit and vegetable processing and preserving industry. The trend is similarly evident in industries such as soft drink and bottled water manufacturing, as well as global cocoa, chocolate, and sugar confectionery manufacturing, forecast to grow by 5.7% and 10.0% respectively during 2008 in response to soaring demand in Chinese and Southeast Asian markets.

Rising food prices over recent years have been linked with social unrest around the world, including rioting in Bangladesh and Mexico, and the Arab Spring.

In 2013 Overseas Development Institute researchers showed that rice has more than doubled in price since 2000, rising by 120% in real terms. This was as a result of shifts in trade policy and restocking by major producers. More fundamental drivers of increased prices are the higher costs of fertiliser, diesel and labour. Parts of Asia see rural wages rise with potential large benefits for the 1.3 billion (2008 estimate) of Asia's poor in reducing the poverty they face. However, this negatively impacts more vulnerable groups who don't share in the economic boom, especially in Asian and African coastal cities. The researchers said the threat means social-protection policies are needed to guard against price shocks. The research proposed that in the longer run, the rises present opportunities to export for Western African farmers with high potential for rice production to replace imports with domestic production.

As Investment

Institutions such as hedge funds, pension funds and investment banks like Barclays Capital, Goldman Sachs and Morgan Stanley have been instrumental in pushing up prices in the last five years, with investment in food commodities rising from $65bn to $126bn (£41bn to £79bn) between 2007 and 2012, contributing to 30-year highs. This has caused price fluctuations which are not strongly related to the actual supply of food, according to the United Nations. Financial institutions now make up 61% of all investment in wheat futures. According to Olivier De Schutter, the UN special rapporteur on food, there was a rush by institutions to enter the food market following George W Bush's Commodities Futures Modernization Act of 2000. De Schutter told the Independent in March 2012: "What we are seeing now is that these financial markets have developed massively with the arrival of these new financial investors, who are purely interested in the short-term monetary gain and are not really interested in the physical thing – they never actually buy the ton of wheat or maize; they only buy a promise to buy or to sell. The result of this financialisation of the commodities market is that the prices of the products respond increasingly to a purely speculative logic. This explains why in very short periods of time we see prices spiking or bubbles exploding, because prices are less and less determined by the real match between supply and demand." In 2011, 450 economists from around the world called on the G20 to regulate the commodities market more.

Some experts have said that speculation has merely aggravated other factors, such as climate change, competition with bio-fuels and overall rising demand. However, some such as Jayati Ghosh, professor of economics at Jawaharlal Nehru University in New Delhi, have pointed out that prices have increased irrespective of supply and demand issues: Ghosh points to world wheat prices, which doubled in the period from June to December 2010, despite there being no fall in global supply.

Famine and Hunger

Food deprivation leads to malnutrition and ultimately starvation. This is often connected with famine, which involves the absence of food in entire communities. This can have a devastating and widespread effect on human health and mortality. Rationing is sometimes used to distribute food in times of shortage, most notably during times of war.

Starvation is a significant international problem. Approximately 815 million people are undernourished, and over 16,000 children die per day from hunger-related causes. Food deprivation is regarded as a deficit need in Maslow's hierarchy of needs and is measured using famine scales.

Food Aid

Food aid can benefit people suffering from a shortage of food. It can be used to improve peoples' lives in the short term, so that a society can increase its standard of living to the point that food aid is no longer required. Conversely, badly managed food aid can create problems by disrupting local markets, depressing crop prices, and discouraging food production. Sometimes a cycle of food aid dependence can develop. Its provision, or threatened withdrawal, is sometimes used as a political tool to influence the policies of the destination country, a strategy known as food politics. Sometimes, food aid provisions will require certain types of food be purchased from certain sellers, and food aid can be misused to enhance the markets of donor countries. International efforts to distribute food to the neediest countries are often coordinated by the World Food Programme.

Safety

Foodborne illness, commonly called "food poisoning", is caused by bacteria, toxins, viruses, parasites, and prions. Roughly 7 million people die of food poisoning each year, with about 10 times as many suffering from a non-fatal version. The two most common factors leading to cases of bacterial foodborne illness are cross-contamination of ready-to-eat food from other uncooked foods and improper temperature control. Less commonly, acute adverse reactions can also occur if chemical contamination of food occurs, for example from improper storage, or use of non-food grade soaps and disinfectants. Food can also be adulterated by a very wide range of articles (known as "foreign bodies") during farming, manufacture, cooking, packaging, distribution, or sale. These foreign bodies can include pests or their droppings, hairs, cigarette butts, wood chips, and all manner of other contaminants. It is possible for certain types of food to become contaminated if stored or presented in an unsafe container, such as a ceramic pot with lead-based glaze.

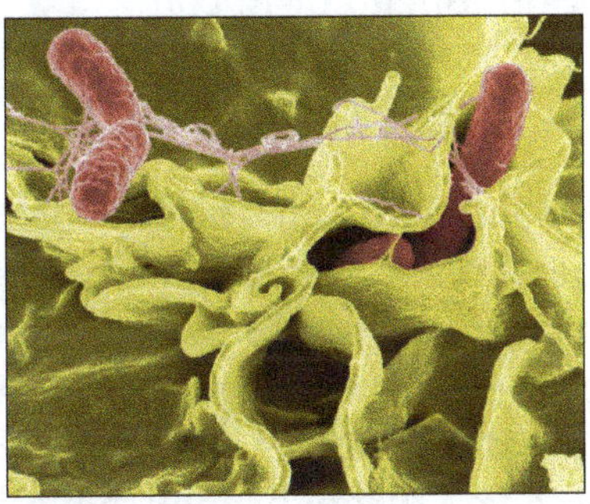

Salmonella bacteria is a common cause of foodborne illness, particularly in undercooked chicken and chicken eggs.

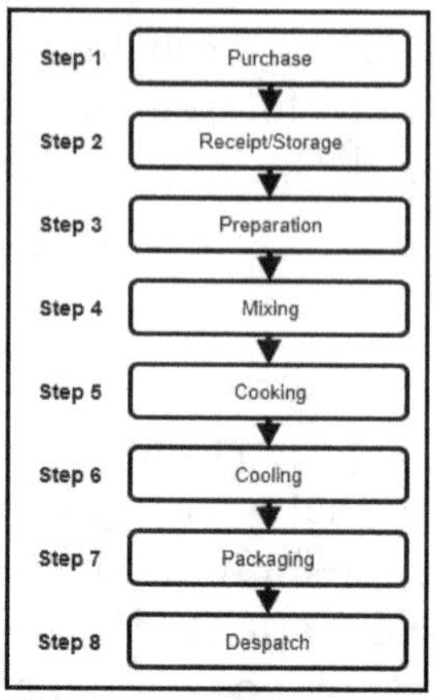

Hazard Analysis and Critical Control Points (HACCP) Flowchart.

Food poisoning has been recognized as a disease since as early as Hippocrates. The sale of rancid, contaminated, or adulterated food was commonplace until the introduction of hygiene, refrigeration, and vermin controls in the 19th century. Discovery of techniques for killing bacteria using heat, and other microbiological studies by scientists such as Louis Pasteur, contributed to the modern sanitation standards that are ubiquitous in developed nations today. This was further underpinned by the work of Justus von Liebig, which led to the development of modern food storage and food preservation methods. In more recent years, a greater understanding of the causes of food-borne illnesses has led to the development of more systematic approaches such as the Hazard Analysis and Critical Control Points (HACCP), which can identify and eliminate many risks.

Recommended measures for ensuring food safety include maintaining a clean preparation area with foods of different types kept separate, ensuring an adequate cooking temperature, and refrigerating foods promptly after cooking.

Foods that spoil easily, such as meats, dairy, and seafood, must be prepared a certain way to avoid contaminating the people for whom they are prepared. As such, the rule of thumb is that cold foods (such as dairy products) should be kept cold and hot foods (such as soup) should be kept hot until storage. Cold meats, such as chicken, that are to be cooked should not be placed at room temperature for thawing, at the risk of dangerous bacterial growth, such as *Salmonella* or *E. coli*.

Allergies

Some people have allergies or sensitivities to foods which are not problematic to most people. This occurs when a person's immune system mistakes a certain food protein for a harmful foreign agent and attacks it. About 2% of adults and 8% of children have a food allergy. The amount of the food substance required to provoke a reaction in a particularly susceptible individual can be quite small. In some instances, traces of food in the air, too minute to be perceived through smell, have been known to provoke lethal reactions in extremely sensitive individuals. Common food allergens are gluten, corn, shellfish (mollusks), peanuts, and soy. Allergens frequently produce symptoms such as diarrhea, rashes, bloating, vomiting, and regurgitation. The digestive complaints usually develop within half an hour of ingesting the allergen.

Rarely, food allergies can lead to a medical emergency, such as anaphylactic shock, hypotension (low blood pressure), and loss of consciousness. An allergen associated with this type of reaction is peanut, although latex products can induce similar reactions. Initial treatment is with epinephrine (adrenaline), often carried by known patients in the form of an Epi-pen or Twinject.

Other Health Issues

Human diet was estimated to cause perhaps around 35% of cancers in a human epidemiological analysis by Richard Doll and Richard Peto in 1981. These cancer may be caused by carcinogens that are present in food naturally or as contaminants. Food contaminated with fungal growth may contain mycotoxins such as aflatoxins which may be found in contaminated corn and peanuts. Other carcinogens identified in food include heterocyclic amines generated in meat when cooked at high temperature, polyaromatic hydrocarbons in charred meat and smoked fish, and nitrosamines generated from nitrites used as food preservatives in cured meat such as bacon.

Anticarcinogens that may help prevent cancer can also be found in many food especially fruits and vegetable. Antioxidants are important groups of compounds that may help remove potentially harmful chemicals. It is however often difficult to identify the specific components in diet that serve to increase or decrease cancer risk since many food, such as beef steak and broccoli, contain low concentrations of both carcinogens and anticarcinogens.

Diet

Other area (Yr 2010) * Africa, sub-Sahara - 2170 kcal/capita/day * N.E. and N. Africa - 3120 kcal/capita/day * South Asia - 2450 kcal/capita/day * East Asia - 3040 kcal/capita/day * Latin America / Caribbean - 2950 kcal/capita/day * Developed countries - 3470 kcal/capita/day.

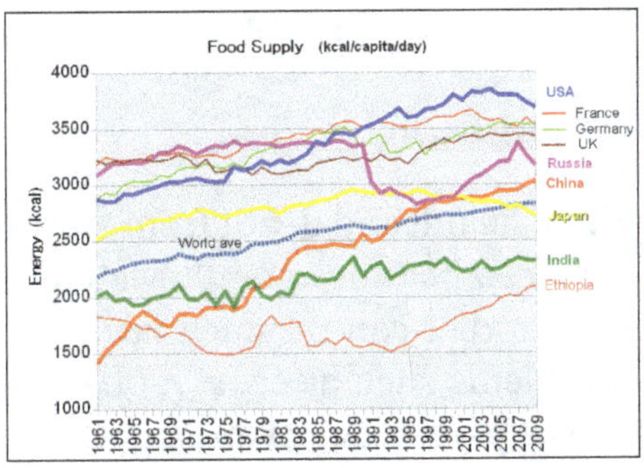

Changes of food supply (by energy).

Cultural and Religious Diets

Dietary habits are the habitual decisions a person or culture makes when choosing what foods to eat. Many cultures hold some food preferences and some food taboos. Dietary choices can also define cultures and play a role in religion. For example, only kosher foods are permitted by Judaism, halal foods by Islam, and in Hinduism beef is restricted. In addition, the dietary choices of different countries or regions have different characteristics. This is highly related to a culture's cuisine.

Diet Deficiencies

Dietary habits play a significant role in the health and mortality of all humans. Imbalances between the consumed fuels and expended energy results in either starvation or excessive reserves of adipose tissue, known as body fat. Poor intake of various vitamins and minerals can lead to diseases that can have far-reaching effects on health. For instance, 30% of the world's population either has, or is at risk for developing, iodine deficiency. It is estimated that at least 3 million children are blind due to vitamin A deficiency. Vitamin C deficiency results in scurvy. Calcium, Vitamin D, and phosphorus are inter-related; the consumption of each may affect the absorption of the others. Kwashiorkor and marasmus are childhood disorders caused by lack of dietary protein.

Moral, Ethical, and Health-conscious Diets

Many individuals limit what foods they eat for reasons of morality, or other habit. For instance, vegetarians choose to forgo food from animal sources to varying degrees. Others choose a healthier diet, avoiding sugars or animal fats and increasing consumption of dietary fiber and antioxidants. Obesity, a serious problem in the western world, leads to higher chances of developing heart disease, diabetes, and many other diseases. More recently, dietary habits have been influenced by the concerns that some people have about possible impacts on health or the environment from genetically modified food. Further concerns about the impact of industrial farming (grains) on animal welfare, human health, and the environment are also having an effect on contemporary human dietary habits. This has led to the emergence of a movement with a preference for organic and local food.

Nutrition and Dietary Problems

Between the extremes of optimal health and death from starvation or malnutrition, there is an array of disease states that can be caused or alleviated by changes in diet. Deficiencies, excesses, and imbalances in diet can produce negative impacts on health, which may lead to various health problems such as scurvy, obesity, or osteoporosis, diabetes, cardiovascular diseases as well as psychological and behavioral problems. The science of nutrition attempts to understand how and why specific dietary aspects influence health.

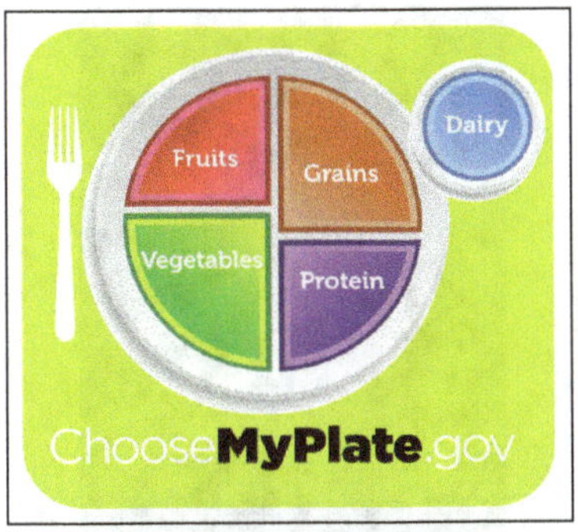

MyPlate replaced MyPyramid as the USDA nutrition guide.

Nutrients in food are grouped into several categories. Macronutrients are fat, protein, and carbohydrates. Micronutrients are the minerals and vitamins. Additionally, food contains water and dietary fiber.

As previously discussed, the body is designed by natural selection to enjoy sweet and fattening foods for evolutionary diets, ideal for hunters and gatherers. Thus, sweet and fattening foods in nature are typically rare and are very pleasurable to eat. In modern times, with advanced technology, enjoyable foods are easily available to consumers. Unfortunately, this promotes obesity in adults and children alike.

Legal Definition

Some countries list a legal definition of food, often referring them with the word *foodstuff*. These countries list food as any item that is to be processed, partially processed, or unprocessed for consumption. The listing of items included as food include any substance intended to be, or reasonably expected to be, ingested by humans. In addition to these foodstuffs, drink, chewing gum, water, or other items processed into said food items are part of the legal definition of food. Items not included in the legal definition of food include animal feed, live animals (unless being prepared for sale in a market), plants prior to harvesting, medicinal products, cosmetics, tobacco and tobacco products, narcotic or psychotropic substances, and residues and contaminants.

Fiber

Fiber or fibre (from the Latin fibra) is a natural or synthetic substance that is significantly longer than it is wide. Fibers are often used in the manufacture of other materials. The strongest engineering materials often incorporate fibers, for example carbon fiber and ultra-high-molecular-weight polyethylene.

Synthetic fibers can often be produced very cheaply and in large amounts compared to natural fibers, but for clothing natural fibers can give some benefits, such as comfort, over their synthetic counterparts.

A bundle of optical fibers.

Natural Fibers

Natural fibers develop or occur in the fiber shape, and include those produced by plants, animals, and geological processes. They can be classified according to their origin:

- Vegetable fibers are generally based on arrangements of cellulose, often with lignin: examples include cotton, hemp, jute, flax, ramie, sisal, bagasse, and banana. Plant fibers are employed in the manufacture of paper and textile (cloth), and dietary fiber is an important component of human nutrition.
- Wood fiber, distinguished from vegetable fiber, is from tree sources. Forms include groundwood, lacebark, thermomechanical pulp (TMP), and bleached or unbleached kraft or sulfite pulps. Kraft and sulfite (also called sulphite) refer to the type of pulping process used to remove the lignin bonding the original wood structure, thus freeing the fibers for use in paper and engineered wood products such as fiberboard.
- Animal fibers consist largely of particular proteins. Instances are silkworm silk, spider silk, sinew, catgut, wool, sea silk and hair such as cashmere wool, mohair and angora, fur such as sheepskin, rabbit, mink, fox, beaver, etc.
- Mineral fibers include the asbestos group. Asbestos is the only naturally occurring long mineral fiber. Six minerals have been classified as "asbestos" including chrysotile of the serpentine class and those belonging to the amphibole class: amosite, crocidolite, tremolite, anthophyllite and actinolite. Short, fiber-like minerals include wollastonite and palygorskite.
- Biological fibers also known as fibrous proteins or protein filaments consist largely of biologically relevant and biologically very important proteins, mutations or other genetic defects can lead to severe diseases. Instances are collagen family of proteins, tendon, muscle proteins like actin, cell proteins like microtubules and many others, spider silk, sinew and hair etc.

Man-made Fibers

Man-made fibers or chemical fibers are fibers whose chemical composition, structure, and properties are significantly modified during the manufacturing process. Man-made fibers consist of regenerated fibers and synthetic fibers.

Semi-synthetic Fibers

Semi-synthetic fibers are made from raw materials with naturally long-chain polymer structure and are only modified and partially degraded by chemical processes, in contrast to completely synthetic fibers such as nylon (polyamide) or dacron (polyester), which the chemist synthesizes from low-molecular weight compounds by polymerization (chain-building) reactions. The earliest semi-synthetic fiber is the cellulose regenerated fiber, rayon. Most semi-synthetic fibers are cellulose regenerated fibers.

Cellulose Regenerated Fibers

Cellulose fibers are a subset of man-made fibers, regenerated from natural cellulose. The cellulose comes from various sources: rayon from tree wood fiber, Modal from beech trees, bamboo fiber from bamboo, seacell from seaweed, etc. In the production of these fibers, the cellulose is reduced to a fairly pure form as a viscous mass and formed into fibers by extrusion through spinnerets. Therefore, the manufacturing process leaves few characteristics distinctive of the natural source material in the finished products.

Some examples are:

- Rayon.
- Bamboo fiber.
- Lyocell, a brand of rayon.
- Modal, using beech trees as input.
- Diacetate fiber.
- Triacetate fiber.

Historically, cellulose diacetate and -triacetate were classified under the term rayon, but are now considered distinct materials.

Synthetic Fibers

Synthetic come entirely from synthetic materials such as petrochemicals, unlike those man-made fibers derived from such natural substances as cellulose or protein.

Fiber classification in reinforced plastics falls into two classes: (i) short fibers, also known as discontinuous fibers, with a general aspect ratio (defined as the ratio of fiber length to diameter) between 20 and 60, and (ii) long fibers, also known as continuous fibers, the general aspect ratio is between 200 and 500.

Metallic Fibers

Metallic fibers can be drawn from ductile metals such as copper, gold or silver and extruded or deposited from more brittle ones, such as nickel, aluminum or iron.

Carbon Fiber

Carbon fibers are often based on oxydized and via pyrolysis carbonized polymers like PAN, but the end product is almost pure carbon.

Silicon Carbide Fiber

Silicon carbide fibers, where the basic polymers are not hydrocarbons but polymers, where about 50% of the carbon atoms are replaced by silicon atoms, so-called poly-carbo-silanes. The pyrolysis yields an amorphous silicon carbide, including mostly other elements like oxygen, titanium, or aluminium, but with mechanical properties very similar to those of carbon fibers.

Fiberglass

Fiberglass, made from specific glass, and optical fiber, made from purified natural quartz, are also man-made fibers that come from natural raw materials, silica fiber, made from sodium silicate (water glass) and basalt fiber made from melted basalt.

Mineral Fibers

Mineral fibers can be particularly strong because they are formed with a low number of surface defects, asbestos is a common one.

Polymer Fibers

- Polymer fibers are a subset of man-made fibers, which are based on synthetic chemicals (often from petrochemical sources) rather than arising from natural materials by a purely physical process. These fibers are made from:
  - Polyamide nylon.
  - PET or PBT polyester.
  - Phenol-formaldehyde (PF).

- Polyvinyl chloride fiber (PVC) vinyon.
- Polyolefins (PP and PE) olefin fiber.
- Acrylic polyesters, pure polyester PAN fibers are used to make carbon fiber by roasting them in a low oxygen environment. Traditional acrylic fiber is used more often as a synthetic replacement for wool. Carbon fibers and PF fibers are noted as two resin-based fibers that are not thermoplastic, most others can be melted.
- Aromatic polyamids (aramids) such as Twaron, Kevlar and Nomex thermally degrade at high temperatures and do not melt. These fibers have strong bonding between polymer chains.
- Polyethylene (PE), eventually with extremely long chains / HMPE (e.g. Dyneema or Spectra).
- Elastomers can even be used, e.g. spandex although urethane fibers are starting to replace spandex technology.
- Polyurethane fiber.
- Elastolefin.

- Coextruded fibers have two distinct polymers forming the fiber, usually as a core-sheath or side-by-side. Coated fibers exist such as nickel-coated to provide static elimination, silver-coated to provide anti-bacterial properties and aluminum-coated to provide RF deflection for radar chaff. Radar chaff is actually a spool of continuous glass tow that has been aluminum coated. An aircraft-mounted high speed cutter chops it up as it spews from a moving aircraft to confuse radar signals.

Microfibers

Microfibers in textiles refer to sub-denier fiber (such as polyester drawn to 0.5 denier). Denier and Dtex are two measurements of fiber yield based on weight and length. If the fiber density is known, you also have a fiber diameter, otherwise it is simpler to measure diameters in micrometers. Microfibers in technical fibers refer to ultra fine fibers (glass or meltblown thermoplastics) often used in filtration. Newer fiber designs include extruding fiber that splits into multiple finer fibers. Most synthetic fibers are round in cross-section, but special designs can be hollow, oval, star-shaped or trilobal. The latter design provides more optically reflective properties. Synthetic textile fibers are often crimped to provide bulk in a woven, non woven or knitted structure. Fiber surfaces can also be dull or bright. Dull surfaces reflect more light while bright tends to transmit light and make the fiber more transparent.

Very short and/or irregular fibers have been called fibrils. Natural cellulose, such as cotton or bleached kraft, show smaller fibrils jutting out and away from the main fiber structure.

Dairy Product

A dairy product or *milk product* is food produced from the milk of mammals, primarily cows, water buffaloes, goats, sheep, yaks, horses, camels, and domestic buffaloes. A facility that processes milk is a dairy or a dairy factory. Dairy products are commonly found in European, Middle Eastern, South Asian, and Central Asian cuisines, but are largely absent from East Asian and Southeast Asian cuisines.

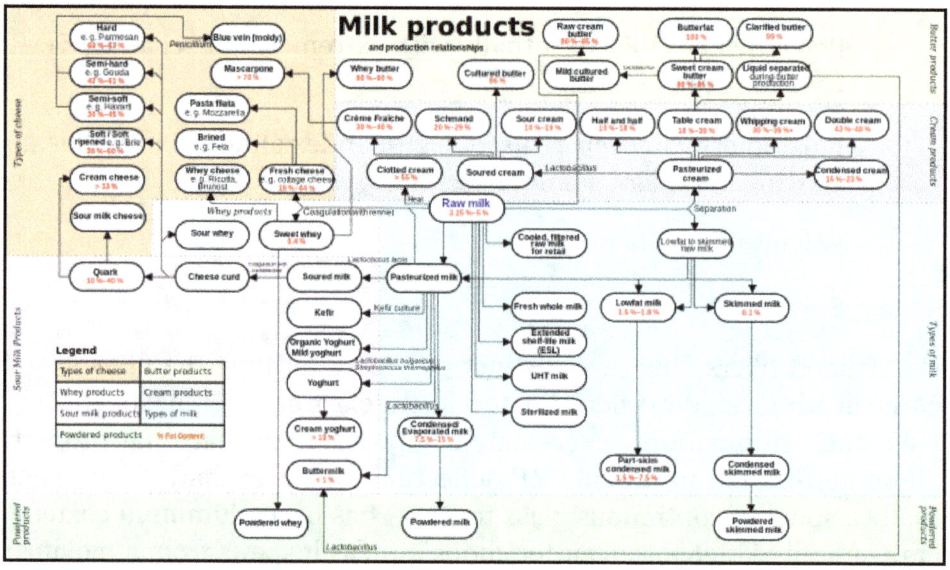

Milk products and production relationships.

Types of Dairy Products

A dairy farm.

The milk products of the Water buffaloes (super carabaos, Philippine Carabao Center).

- Milk after optional homogenization, pasteurization, in several grades after standardization of the fat level, and possible addition of the bacteria *Streptococcus lactis* and *Leuconostoc citrovorum*:
    - *Crème fraîche*, slightly fermented cream:
        - Clotted cream, thick, spoonable cream made by heating milk.
        - Single cream, double cream and whipping cream.
        - *Smetana*, Central and Eastern European variety of sour cream.
    - Cultured milk resembling buttermilk, but uses different yeast and bacterial cultures.
    - Kefir, fermented milk drink from the Northern Caucasus.
    - *Kumis/Airag*, slightly fermented mares' milk popular in Central Asia.
    - Powdered milk (or milk powder), produced by removing the water from (usually skim) milk:
        - Whole milk products.
        - Buttermilk products.
        - Skim milk.
        - Whey products.
        - High milk-fat and nutritional products (for infant formulas).
        - Cultured and confectionery products.
    - Condensed milk, milk which has been concentrated by evaporation, with sugar added for reduced process time and longer life in an opened can.
    - *Khoa*, milk which has been completely concentrated by evaporation, used in Indian cuisine including gulab jamun, peda, etc.).

- Evaporated milk, (less concentrated than condensed) milk without added sugar.
- Ricotta, acidified whey, reduced in volume.
- Infant formula, dried milk powder with specific additives for feeding human infants.
- Baked milk, a variety of boiled milk that has been particularly popular in Russia.

- Butter, mostly milk fat, produced by churning cream:
  - Buttermilk, the liquid left over after producing butter from cream, often dried as livestock feed.
  - *Ghee*, clarified butter, by gentle heating of butter and removal of the solid matter.
  - *Smen*, a fermented, clarified butter used in Moroccan cooking.
  - Anhydrous milkfat (clarified butter).
- Cheese, produced by coagulating milk, separating from whey and letting it ripen, generally with bacteria and sometimes also with certain molds:
  - Curds, the soft, curdled part of milk (or skim milk) used to make cheese.
  - *Paneer*.
  - Whey, the liquid drained from curds and used for further processing or as a livestock feed.
  - Cottage cheese.
  - Quark.
  - Cream cheese, produced by the addition of cream to milk and then curdled to form a rich curd or cheese.
  - *Fromage frais.*
- Casein are:
  - Caseinates, sodium or calcium salts of casein.
  - Milk protein concentrates and isolates.
  - Whey protein concentrates and isolates, reduced lactose whey.

- Hydrolysates, milk treated with proteolytic enzymes to alter functionality.
- Mineral concentrates, byproduct of demineralizing whey.

- Yogurt, milk fermented by *Streptococcus salivarius* ssp. *thermophilus* and *Lactobacillus delbrueckii* ssp. *bulgaricus* sometimes with additional bacteria, such as *Lactobacillus acidophilus*:
  - *Ayran*.
  - *Lassi*, Indian subcontinent.
  - *Leben*.
- Clabber, milk naturally fermented to a yogurt-like state.
- Gelato, slowly frozen milk and water, lesser fat than ice cream.
- Ice cream, slowly frozen cream, milk, flavors and emulsifying additives (dairy ice cream):
  - Ice milk, low-fat version of ice cream.
  - Frozen custard.
  - Frozen yogurt, yogurt with emulsifiers.
- Other:
  - *Viili*.
  - *Kajmak*.
  - *Filmjölk*.
  - *Piimä*.
  - *Vla*.
  - *Dulce de leche*.
  - *Skyr*.
  - Junket, milk solidified with rennet.

Health

Dairy products can cause health issues for individuals who have lactose intolerance or a milk allergy.

Additionally dairy products including cheese, ice cream, milk, butter, and yogurt can contribute significant amounts of cholesterol and saturated fat to the diet. Diets high in fat and especially in saturated fat can increase the risk of heart disease and can cause other serious health problems. However, it has been shown that there is no connection between dairy consumption (excluding butter) and cardiovascular disease, even though dairy tends to be higher in saturated fats.

Avoidance

Some groups avoid dairy products for non-health related reasons:

- Religious - Some religions restrict or do not allow for the consumption of dairy products. For example, some scholars of Jainism advocate not consuming any dairy products because dairy is perceived to involve violence against cows. Strict Judaism requires that meat and dairy products not be served at the same meal, served or cooked in the same utensils, or stored together, as prescribed in Deuteronomy 14:21.
- Ethical - Veganism is the avoidance of all animal products, including dairy products, most often due to the ethics regarding how dairy products are produced. The ethical reasons for avoiding dairy include how dairy is produced, how the animals are handled, and the environmental effect of dairy production.

Egg

Eggs are laid by female animals of many different species, including birds, reptiles, amphibians, mammals, and fish, and have been eaten by humans for thousands of years. Bird and reptile eggs consist of a protective eggshell, albumen (egg white), and vitellus (egg yolk), contained within various thin membranes. The most popular choice for egg consumption are chicken eggs. Other popular choices for egg consumption are duck, quail, roe, and caviar.

Eggs of various birds, a reptile, various cartilaginous fish, a cuttlefish and various butterflies and moths. (Click on image for key).

Egg yolks and whole eggs store significant amounts of protein and choline, and are widely used in cookery. Due to their protein content, the United States Department of Agriculture categorizes eggs as *Meats* within the Food Guide Pyramid. Despite the nutritional value of eggs, there are some potential health issues arising from egg quality, storage, and individual allergies.

Chickens and other egg-laying creatures are widely kept throughout the world, and mass production of chicken eggs is a global industry. In 2009, an estimated 62.1 million metric tons of eggs were produced worldwide from a total laying flock of approximately 6.4 billion hens. There are issues of regional variation in demand and expectation, as well as current debates concerning methods of mass production. The European Union recently banned battery husbandry of chickens.

History

Bird eggs have been valuable foodstuffs since prehistory, in both hunting societies and more recent cultures where birds were domesticated. The chicken was probably domesticated for its eggs from jungle fowl native to tropical and subtropical Southeast Asia and India before 7500 BCE. Chickens were brought to Sumer and Egypt by 1500 BCE, and arrived in Greece around 800 BCE, where the quail had been the primary source of eggs. In Thebes, Egypt, the tomb of Haremhab, built about 1420 BCE, shows a depiction of a man carrying bowls of ostrich eggs and other large eggs, presumably those of the pelican, as offerings. In ancient Rome, eggs were preserved using a number of methods, and meals often started with an egg course. The Romans crushed the shells in their plates to prevent evil spirits from hiding there. In the Middle Ages, eggs were forbidden during Lent because of their richness. The word mayonnaise possibly was derived from *moyeu*, the medieval French word for the yolk, meaning center or hub.

Egg scrambled with acidic fruit juices were popular in France in the 17th century; this may have been the origin of lemon curd.

The dried egg industry developed in the 19th century, before the rise of the frozen egg industry. In 1878, a company in St. Louis, Missouri started to transform egg yolk and white into a light-brown, meal-like substance by using a drying process. The production of dried eggs significantly expanded during World War II, for use by the United States Armed Forces and its allies.

In 1911, the egg carton was invented by Joseph Coyle in Smithers, British Columbia, to solve a dispute about broken eggs between a farmer in Bulkley Valley and the owner of the Aldermere Hotel. Early egg cartons were made of paper.

Varieties

Bird eggs are a common food and one of the most versatile ingredients used in cooking. They are important in many branches of the modern food industry.

Quail eggs (upper left), chicken egg (lower left) and ostrich egg (right).

The most commonly used bird eggs are those from the chicken. Duck and goose eggs, and smaller eggs, such as quail eggs, are occasionally used as a gourmet ingredient in western countries. Eggs are a common everyday food in many parts of Asia, such as China and Thailand, with Asian production providing 59% of the world total in 2013.

The largest bird eggs, from ostriches tend to be used only as special luxury food. Gull eggs are considered a delicacy in England, as well as in some Scandinavian countries, particularly in Norway. In some African countries, guineafowl eggs are commonly seen in marketplaces, especially in the spring of each year. Pheasant eggs and emu eggs are edible, but less widely available. Sometimes they are obtainable from farmers, poulterers, or luxury grocery stores. In many countries, wild birds' eggs are protected by laws which prohibit collecting or selling them, or permit collection only during specific periods of the year.

Global production of chicken eggs (in shell) (2013, in millions of tonnes)	
China	24.8
USA	5.6
India	3.8
Japan	2.5
World	**68.3**
Source: United Nations, Food and Agriculture Organization, FAOSTAT	

Production

In 2013, world production of chicken eggs was 68.3 million tonnes, with China contributing 37% of this total.

Anatomy and Characteristics

The shape of an egg resembles a prolate spheroid with one end larger than the other, with cylindrical symmetry along the long axis.

A raw chicken egg within its membrane, the shell removed by soaking in vinegar.

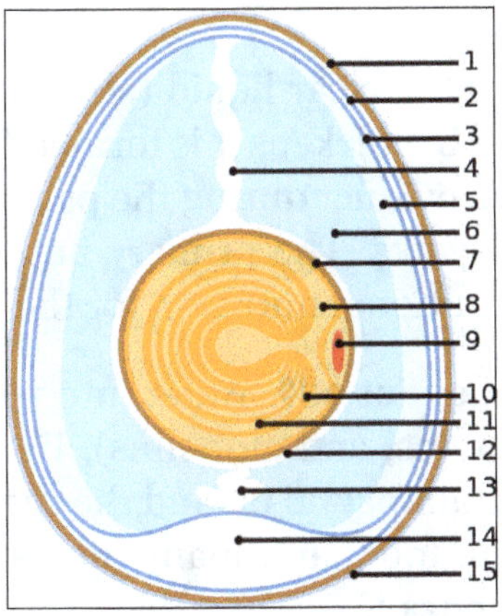

Schematic of a chicken egg:
1. Eggshell, 2. Outer membrane, 3. Inner membrane, 4. Chalaza, 5. Exterior albumen, 6. Middle albumen, 7. Vitelline membrane, 8. Nucleus of pander, 9. Germinal disc (nucleus), 10. Yellow yolk, 11. White yolk 12. Internal albumen, 13. Chalaza, 14. Air cell, 15. Cuticula.

An egg is surrounded by a thin, hard shell. Inside, the egg yolk is suspended in the egg white by one or two spiral bands of tissue called the chalazae.

Air Cell

The larger end of the egg contains the air cell that forms when the contents of the egg cool down and contract after it is laid. Chicken eggs are graded according to the size of this air cell, measured during candling. A very fresh egg has a small air cell and receives a grade of AA. As the size of the air cell increases and the quality of the egg decreases, the grade moves from AA to A to B. This provides a way of testing the age of an egg: as the air cell increases in size due to air being drawn through pores in the shell as water is lost, the egg becomes less dense and the larger end of the egg will rise to increasingly shallower depths when the egg is placed in a bowl of water. A very old egg will actually float in the water and should not be eaten.

Shell

Egg shell color is caused by pigment deposition during egg formation in the oviduct and can vary according to species and breed, from the more common white or brown to pink or speckled blue-green. In general, chicken breeds with white ear lobes lay white eggs, whereas chickens with red ear lobes lay brown eggs. Although there is no significant link between shell color and nutritional value, there is often a cultural preference for one color over another.

Membrane

The membrane is a clear film lining the eggshell, visible when one peels a boiled egg. Eggshell membrane is primarily composed of fibrous proteins such as collagen type I.

White

White is the common name for the clear liquid (also called the albumen or the glair/glaire) contained within an egg. In chickens it is formed from the layers of secretions of the anterior section of the hen's oviduct during the passage of the egg. It forms around either fertilized or unfertilized yolks. The primary natural purpose of egg white is to protect the yolk and provide additional nutrition for the growth of the embryo.

Egg white consists primarily of about 90% water into which is dissolved 10% proteins (including albumins, mucoproteins, and globulins). Unlike the yolk, which is high in lipids (fats), egg white contains almost no fat and the carbohydrate content is less than 1%. Egg white has many uses in food, and many others, including the preparation of vaccines such as those for influenza.

Yolk

The yolk in a newly laid egg is round and firm. As the yolk ages, it absorbs water from the albumen, which increases its size and causes it to stretch and weaken the vitelline membrane (the clear casing enclosing the yolk). The resulting effect is a flattened and enlarged yolk shape.

Yolk color is dependent on the diet of the hen; if the diet contains yellow/orange plant pigments known as xanthophylls, then they are deposited in the yolk, coloring it. Lutein is the most abundant pigment in egg yolk. A colorless diet can produce an almost colorless yolk. Yolk color is, for example, enhanced if the diet includes products such as yellow corn and marigold petals. In the US, the use of artificial color additives is forbidden.

Abnormalities

Shell-less or thin-shelled eggs can be caused by egg drop syndrome.

Culinary Properties

Types of Dishes

Chicken eggs are widely used in many types of dishes, both sweet and savory, including many baked goods. Some of the most common preparation methods include scrambled, fried, hard-boiled, soft-boiled, omelettes and pickled. They can also be eaten raw, though this is not recommended for people who may be especially susceptible to salmonellosis, such as the elderly, the infirm, or pregnant women. In addition, the protein in raw eggs is only 51% bioavailable, whereas that of a cooked egg is nearer 91% bio-available, meaning the protein of cooked eggs is nearly twice as absorbable as the protein from raw eggs.

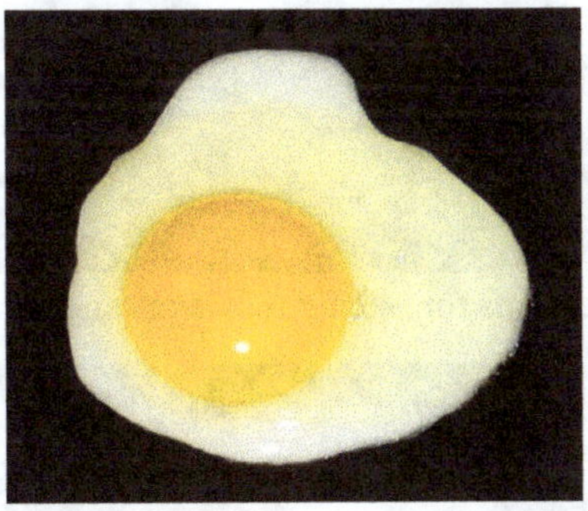

A fried chicken egg, "sunny side up".

As an ingredient, egg yolks are an important emulsifier in the kitchen, and are also used as a thickener in custards.

Soft-boiled quail eggs, with potato galettes.

The albumen, or egg white, contains protein, but little or no fat, and can be used in cooking separately from the yolk. The proteins in egg white allow it to form foams and aerated dishes. Egg whites may be aerated or whipped to a light, fluffy consistency, and are often used in desserts such as meringues and mousse.

Ground egg shells are sometimes used as a food additive to deliver calcium. Every part of an egg is edible, although the eggshell is generally discarded. Some recipes call for immature or unlaid eggs, which are harvested after the hen is slaughtered or cooked while still inside the chicken.

Cooking

Eggs contain multiple proteins which gel at different temperatures within the yolk and the white, and the temperature determines the gelling time. Egg yolk begins to gelify, or solidify, when it reaches temperatures between about 60 and 70 °C (140 and 158 °F). Egg white gels at slightly higher temperatures, about 60 to 80 °C (140 to 176 °F)- the white contains ovalbumin that sets at the highest temperature. However, in practice, in many cooking processes the white gels first because it is exposed to higher temperatures for longer.

A few tips for baking with eggs and cakes.

half boiled egg dish.

Salmonella is killed instantly at 71 °C (160 °F), but is also killed from 54.5 °C (130.1 °F) if held there for sufficiently long time periods. To avoid the issue of salmonella, eggs can be pasteurized in-shell at 57 °C (135 °F) for an hour and 15 minutes. Although the white is slightly milkier, the eggs can be used in normal ways. Whipping for meringue takes significantly longer, but the final volume is virtually the same.

If a boiled egg is overcooked, a greenish ring sometimes appears around egg yolk due to the iron and sulfur compounds in the egg. It can also occur with an abundance of iron in the cooking water. The green ring does not affect the egg's taste; overcooking, however, harms the quality of the protein. Chilling the egg for a few minutes in cold water until it is completely cooled may prevent the greenish ring from forming on the surface of the yolk.

Flavor Variations

Although the age of the egg and the conditions of its storage have a greater influence, the bird's diet does affect the flavor of the egg. For example, when a brown-egg chicken breed eats rapeseed or soy meals, its intestinal microbes metabolize them into fishy-smelling triethylamine, which ends up in the egg. The unpredictable diet of free-range hens will produce unpredictable eggs. Duck eggs tend to have a flavor distinct from, but still resembling chicken eggs.

A batch of tea eggs with shell intact soaking in a brew of spices and tea.

Eggs can also be soaked in mixtures to absorb flavor. Tea eggs are steeped in a brew from a mixture of various spices, soy sauce, and black tea leaves to give flavor.

Storage

Careful storage of edible eggs is extremely important, as an improperly handled egg can contain elevated levels of *Salmonella* bacteria that can cause severe food poisoning. In the US, eggs are washed, and this cleans the shell, but erodes the cuticle. The USDA thus recommends refrigerating eggs to prevent the growth of *Salmonella*.

Refrigeration also preserves the taste and texture. However, intact eggs can be left unrefrigerated for several months without spoiling. In Europe, eggs are not usually washed, and the shells are dirtier, however the cuticle is undamaged, and they do not require refrigeration. In the UK in particular, hens are immunized against salmonella, and the eggs are generally safe for 21 days.

Preservation

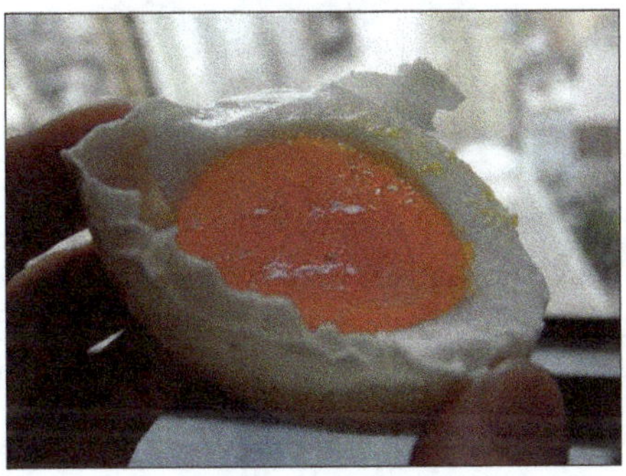

Salted duck egg.

The simplest method to preserve an egg is to treat it with salt. Salt draws water out of bacteria and molds, which prevents their growth. The Chinese salted duck egg is made by immersing duck eggs in brine, or coating them individually with a paste of salt and mud or clay. The eggs stop absorbing salt after about a month, having reached osmotic equilibrium. Their yolks take on an orange-red color and solidify, but the white remains somewhat liquid. They are often boiled before consumption, and are often served with rice congee.

Pickled egg, colored with beetroot juice.

Another method is to make pickled eggs, by boiling them first and immersing them in a mixture of vinegar, salt and spices, such as ginger or allspice. Frequently, beetroot juice is added to impart a red color to the eggs. If the eggs are immersed in it for a few hours, the distinct red, white, and yellow colors can be seen when the eggs are sliced. If marinated for several days or more, the red color will reach the yolk. If the eggs are marinated in the mixture for several weeks or more, the vinegar will dissolve much of the shell's calcium carbonate and penetrate the egg, making it acidic enough to inhibit the growth of bacteria and molds. Pickled eggs made this way will generally keep for a year or more without refrigeration.

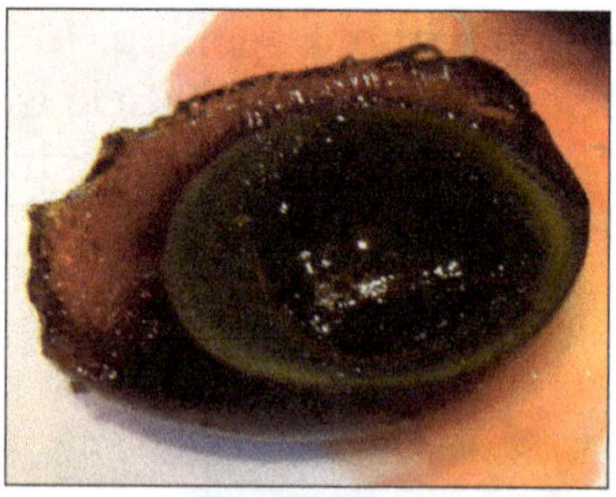

Century egg.

A century egg or hundred-year-old egg is preserved by coating an egg in a mixture of clay, wood ash, salt, lime, and rice straw for several weeks to several months, depending on the method of processing. After the process is completed, the yolk becomes a dark green, cream-like substance with a strong odor of sulfur and ammonia, while the white becomes a dark brown, transparent jelly with a comparatively mild, distinct flavor. The transforming agent in a century egg is its alkaline material, which gradually raises the pH of the egg from around 9 to 12 or more. This chemical process breaks down some of the complex, flavorless proteins and fats of the yolk into simpler, flavorful ones, which in some way may be thought of as an "inorganic" version of fermentation.

Cooking Substitutes

For those who do not consume eggs, alternatives used in baking include other rising agents or binding materials, such as ground flax seeds or potato starch flour. Tofu can also act as a partial binding agent, since it is high in lecithin due to its soy content. Applesauce can be used, as well as arrowroot and banana. Extracted soybean lecithin, in turn, is often used in packaged foods as an inexpensive substitute for egg-derived lecithin. Chickpea brine, also known as aquafaba, can replace egg whites in desserts like meringues and mousses.

Other egg substitutes are made from just the white of the egg for those who worry about the high cholesterol and fat content in eggs. These products usually have added vitamins and minerals, as well as vegetable-based emulsifiers and thickeners such as xanthan gum or guar gum. These allow the product to maintain the nutrition and several culinary properties of real eggs, making possible foods such as Hollandaise sauce, custard, mayonnaise, and most baked goods with these substitutes.

Nutritional Value

Chicken eggs, the most commonly eaten eggs, provide 155 calories (kcal) of food energy and 12.6 g of protein in a 100 gram serving.

Eggs (boiled) supply several vitamins and minerals as significant amounts of the Daily Value (DV), including vitamin A (19% DV), riboflavin (42% DV), pantothenic acid (28% DV), vitamin B_{12} (46% DV), choline (60% DV), phosphorus (25% DV), zinc (11% DV) and vitamin D (15% DV) (table per 100 gram serving of a hard-boiled egg).

A yolk contains more than two-thirds of the recommended daily intake of 300 mg of cholesterol (table).

The diet of laying hens can affect the nutritional quality of eggs. For instance, chicken eggs that are especially high in omega-3 fatty acids are produced by feeding hens a diet containing polyunsaturated fats from sources like fish oil, chia seeds or flaxseeds. Pasture-raised free-range hens, which forage for their own food, also produce eggs that are relatively enriched in omega-3 fatty acids compared to cage-raised chickens.

A 2010 USDA study determined there were no significant differences of macronutrients in various chicken eggs.

Cooked eggs are easier to digest, as well as having a lower risk of salmonellosis.

Health Effects

Cholesterol And Fat

More than half the calories found in eggs come from the fat in the yolk; 50 grams of chicken egg (the contents of an egg just large enough to be classified as "large" in the US, but "medium" in Europe) contains approximately 5 grams of fat. People on a low-cholesterol diet may need to reduce egg consumption; however, only 27% of the fat in egg is saturated fat (palmitic, stearic and myristic acids). The egg white consists primarily of water (87%) and protein (13%) and contains no cholesterol and little, if any, fat.

There is debate over whether egg yolk presents a health risk. Some research suggests dietary cholesterol increases the ratio of total to HDL cholesterol and, therefore, adversely affects the body's cholesterol profile; whereas other studies show that moderate consumption of eggs, up to one a day, does not appear to increase heart disease risk in healthy individuals. Harold McGee argues that the cholesterol in the yolk is not what causes a problem, because fat (in particular, saturated) is much more likely to raise cholesterol levels than the actual consumption of cholesterol.

Type 2 Diabetes

Studies have shown conflicting results about a possible connection between egg consumption and type two diabetes. A 1999 prospective study of over 117,000 people by the Harvard School of Public Health concluded, in part, that "The apparent increased risk of CHD associated with higher egg consumption among diabetic participants warrants further research." A 2008 study by the Physicians' Health Study I (1982–2007) and the Women's Health Study (1992–2007) determined the "data suggest that high levels of egg consumption (daily) are associated with an increased risk of type 2 diabetes." However, a study published in 2010 found no link between egg consumption and type 2 diabetes. A meta-analysis from 2013 found that eating 4 eggs per week increased the risk of diabetes by 29%. Another meta-analysis from 2013 also supported the idea that egg consumption may lead to an increased incidence of type two diabetes.

Cardiovascular Risk

Eggs are one of the largest sources of phosphatidylcholine (lecithin) in the human diet. A study published in the scientific journal Nature showed that dietary phosphatidylcholine is digested by bacteria in the gut and eventually converted into the compound TMAO, a compound linked with increased heart disease.

The 1999 Harvard School of Public Health study of 37,851 men and 80,082 women concluded that its "findings suggest that consumption of up to 1 egg per day is unlikely to have substantial overall impact on the risk of CHD or stroke among healthy men and women." In a study of 4,000 people, scientists found that eating eggs increased blood levels of a metabolite promoting atherosclerosis, TMAO, and that this in turn caused significantly higher risk of heart attack and stroke after three years of follow-up.

A 2007 study of nearly 10,000 adults demonstrated no correlation between moderate (six per week) egg consumption and cardiovascular disease or strokes, except in the subpopulation of diabetic patients who presented an increased risk of coronary artery disease. One potential alternative explanation for the null finding is that background dietary cholesterol may be so high in the usual Western diet that adding somewhat more has little further effect on blood cholesterol. Other research supports the idea that a high egg intake increases cardiovascular risk in diabetic patients. A 2009 prospective cohort study of over 21,000 individuals suggests that "egg consumption up to 6/week has no major effect on the risk of cardiovascular disease and mortality and that consumption of 7+/week is associated with a modest increased risk of total mortality" in males, whereas among males with diabetes, "any egg consumption is associated with an increased risk of all-cause mortality and there was suggestive evidence for an increased risk of myocardial infarction and stroke". A 2013 meta-analysis found no association between egg consumption and heart disease or stroke. A 2013 systematic review and meta-analysis found no association between egg consumption and cardiovascular disease or cardiovascular disease mortality, but did find egg consumption more than once daily increased cardiovascular disease risk 1.69-fold in those with type 2 diabetes mellitus compared to type 2 diabetics who ate less than 1 egg per week. Another 2013 meta-analysis found that eating 4 eggs per week increased the risk of cardiovascular disease by 6%.

Contamination

A health issue associated with eggs is contamination by pathogenic bacteria, such as *Salmonella enteritidis*. Contamination of eggs exiting a female bird via the cloaca may also occur with other members of the *Salmonella* genus, so care must be taken to prevent the egg shell from becoming contaminated with fecal matter. In commercial practice in the US, eggs are quickly washed with a sanitizing solution within minutes of being laid. The risk of infection from raw or undercooked eggs is dependent in part upon the sanitary conditions under which the hens are kept.

Egg cleaning on a farm in Norway.

Health experts advise people to refrigerate washed eggs, use them within two weeks, cook them thoroughly, and never consume raw eggs. As with meat, containers and surfaces that have been used to process raw eggs should not come in contact with ready-to-eat food.

A study by the U.S. Department of Agriculture in 2002 (Risk Analysis April 2002 22(2):203-18) suggests the problem is not as prevalent as once thought. It showed that of the 69 billion eggs produced annually, only 2.3 million are contaminated with *Salmonella*—equivalent to just one in every 30,000 eggs—thus showing *Salmonella* infection is quite rarely induced by eggs. However, this has not been the case in other countries, where *Salmonella enteritidis* and *Salmonella typhimurium* infections due to egg consumptions are major concerns. Egg shells act as hermetic seals that guard against bacteria entering, but this seal can be broken through improper handling or if laid by unhealthy chickens. Most forms of contamination enter through such weaknesses in the shell. In the UK, the British Egg Industry Council award the lions stamp to eggs that, among other things, come from hens that have been vaccinated against *Salmonella*.

Food Allergy

One of the most common food allergies in infants is eggs. Infants usually have the opportunity to grow out of this allergy during childhood, if exposure is minimized. Allergic reactions against egg white are more common than reactions against egg yolks.

In addition to true allergic reactions, some people experience a food intolerance to egg whites.

Food labeling practices in most developed countries now include eggs, egg products and the processing of foods on equipment that also process foods containing eggs in a special allergen alert section of the ingredients on the labels.

Farming Issues

Most commercially farmed chicken eggs intended for human consumption are unfertilized, since the laying hens are kept without roosters. Fertile eggs can be eaten, with a little nutritional difference to the unfertilized. Fertile eggs will not contain a developed embryo, as refrigeration temperatures inhibit cellular growth for an extended time. Sometimes an embryo is allowed to develop but eaten before hatching as with balut.

Eggs for sale at a grocery store.

White and brown eggs in an egg crate.

Grading by Quality and Size

The US Department of Agriculture grades eggs by the interior quality of the egg and the appearance and condition of the egg shell. Eggs of any quality grade may differ in weight (size).

- U.S. Grade AA:
    - Eggs have whites that are thick and firm; yolks that are high, round, and practically free from defects; and clean, unbroken shells.
    - Grade AA and Grade A eggs are best for frying and poaching, where appearance is important.
- U.S. Grade A:
    - Eggs have characteristics of Grade AA eggs except the whites are "reasonably" firm.
    - This is the quality most often sold in stores.
- U.S. Grade B:
    - Eggs have whites that may be thinner and yolks that may be wider and flatter than eggs of higher grades. The shells must be unbroken but may show slight stains.
    - This quality is seldom found in retail stores because they are usually used to make liquid, frozen, and dried egg products, as well as other egg-containing products.

In Australia and the European Union, eggs are graded by the hen farming method, free range, battery caged, etc.

Chicken eggs are also graded by size for the purpose of sales. Some maxi eggs can have double-yolk and some farms separate out double-yolk eggs for special sale.

Comparison of an egg and a maxi egg with a double-yolk - Closed (1/2).

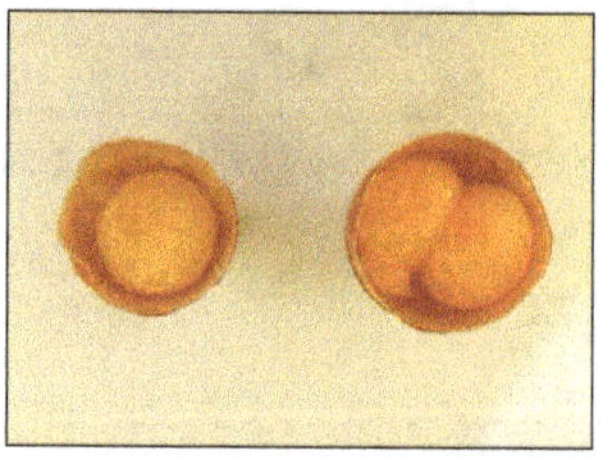

Comparison of an egg and a maxi egg with a double-yolk - Opened (2/2).

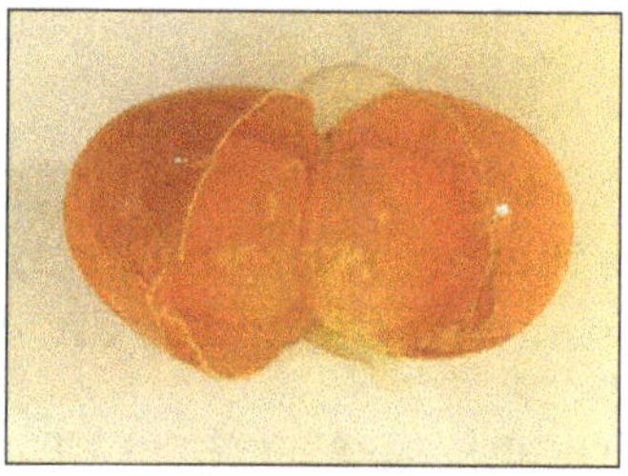

Double-yolk egg - Opened.

Washing and Refrigeration

In North America, the legislation requires eggs to be washed and refrigerated before being sold to consumers. This is to remove natural farm contaminants present in the cleanest farms and to prevent the growth of bacteria. In Europe legislation requires the opposite. Washing removes the natural protective cuticle on the egg and refrigeration causes condensation which may promote bacteria growth.

Color of Eggshell

Although egg color is a largely cosmetic issue, with no effect on egg quality or taste, it is a major issue in production due to regional and national preferences for specific colors, and the results of such preferences on demand. For example, in most regions of the United States, chicken eggs are generally white. In some parts of the northeast of that country, particularly New England, where a television jingle for years proclaimed "brown eggs are local eggs, and local eggs are fresh!", brown eggs are more common. Local chicken breeds, including the Rhode Island Red, lay brown eggs. Brown eggs are also preferred in Costa Rica, Ireland, France, and the United Kingdom. In Brazil and Poland, white chicken eggs are generally regarded as industrial, and brown or red-dish ones are preferred. Small farms and smallholdings, particularly in economically advanced nations, may sell eggs of widely varying colors and sizes, with combinations of white, brown, speckled (red), green, and blue (as laid by certain breeds, including araucanas, heritage skyline and cream leg bar) eggs in the same box or carton, while the supermarkets at the same time sell mostly eggs from the larger producers, of the color preferred in that nation or region.

White, speckled (red), and brown chicken eggs.

Very dark brown eggs of Marans, a French breed of chicken.

These cultural trends have been observed for many years. *The New York Times* reported during the Second World War that housewives in Boston preferred brown eggs and those in New York preferred white eggs. In February 1976, the New Scientist magazine, in discussing issues of chicken egg color, stated "Housewives are particularly fussy about the colour of their eggs, preferring even to pay more for brown eggs although white eggs are just as good". As a result of these trends, brown eggs are usually more expensive to purchase in regions where white eggs are considered 'normal', due to lower production. In France and the United Kingdom, it is very difficult to buy white eggs, with most supermarkets supplying only the more popular brown eggs. By direct contrast, in Egypt it is very hard to source brown eggs, as demand is almost entirely for white ones, with the country's largest supplier describing white eggs as "table eggs" and packaging brown eggs for export.

Research conducted by a France institute in the 1970s demonstrated blue chicken eggs from the Chilean araucana fowl can be stronger and more resilient to breakage.

Research at Nihon University, Japan in 1990 revealed a number of different issues were important to Japanese housewives when deciding which eggs to buy; however, color was a distinct factor, with most Japanese housewives preferring the white color.

Egg producers carefully consider cultural issues, as well as commercial ones, when selecting the breed or breeds of chicken used for production, as egg color varies between breeds. Among producers and breeders, brown eggs are often referred to as "tinted", while the speckled eggs preferred by some consumers are often referred to as being "red" in color.

Living Conditions of Birds

Commercial factory farming operations often involve raising the hens in small, crowded cages, preventing the chickens from engaging in natural behaviors, such as wing-flapping, dust-bathing, scratching, pecking, perching, and nest-building. Such restrictions can lead to pacing and escape behavior.

Laying hens in battery cages.

Many hens confined to battery cages, and some raised in cage-free conditions, are debeaked to prevent them from harming each other and cannibalism. According to critics of the practice, this can cause hens severe pain to the point where some may refuse to eat and starve to death. Some hens may be force molted to increase egg quality and production level after the molting. Molting can be induced by extended feed withdrawal, water withdrawal or controlled lighting programs.

Laying hens are often slaughtered between 100 and 130 weeks of age, when their egg productivity starts to decline. Due to modern selective breeding, laying hen strains differ from meat production strains. As male birds of the laying strain do not lay eggs and are not suitable for meat production, they are generally killed soon after they hatch.

Free-range eggs are considered by some advocates to be an acceptable substitute to factory-farmed eggs. Free-range laying hens are given outdoor access instead of being contained in crowded cages. Questions on the actual living conditions of free-range hens have been raised in the United States of America, as there is no legal definition or regulations for eggs labeled as free-range in that country.

In the United States, increased public concern for animal welfare has pushed various egg producers to promote eggs under a variety of standards. The most widespread standard in use is determined by United Egg Producers through their voluntary program of certification. The United Egg Producers program includes guidelines regarding housing, food, water, air, living space, beak trimming, molting, handling, and transportation, however, opponents such as The Humane Society have alleged UEP Certification is misleading and allows a significant amount of unchecked animal cruelty. Other standards include "Cage Free", "Natural", "Certified Humane", and "Certified Organic". Of these standards, "Certified Humane", which carries requirements for stocking density and cage-free keeping and so on, and "Certified Organic", which requires hens to have outdoor access and be fed only organic vegetarian feed and so on, are the most stringent.

Effective 1 January 2012, the European Union banned conventional battery cages for egg-laying hens, as outlined in EU Directive 1999/74/EC. The EU permits the use of "enriched" furnished cages that must meet certain space and amenity requirements. Egg producers in many member states have objected to the new quality standards while in some countries even furnished cages and family cages are subject to be banned as well. The production standard of the eggs is visible on the mandatory egg marking where the EU egg code begins with 3 for caged chicken to 1 for free-range eggs and 0 for organic egg production.

Killing of Male Chicks

In battery cage and free-range egg production, unwanted male chicks are killed at birth during the process of securing a further generation of egg-laying hens.

Cultural Significance

A popular Easter tradition in some parts of the world is a decoration of hard-boiled eggs (usually by dying, but often by spray-painting). Adults often hide the eggs for children to find, an activity known as an Easter egg hunt. A similar tradition of egg painting exists in areas of the world influenced by the culture of Persia. Before the spring equinox in the Persian New Year tradition (called *Norouz*), each family member decorates a hard-boiled egg and sets them together in a bowl.

Hanácké kraslice, Easter eggs from the Haná region, the Czech Republic.

The tradition of a dancing egg is held during the feast of Corpus Christi in Barcelona and other Catalan cities since the 16th century. It consists of an emptied egg, positioned over the water jet from a fountain, which starts turning without falling.

Although a food item, eggs are sometimes thrown at houses, cars, or people. This act, known commonly as "egging" in the various English-speaking countries, is a minor form of vandalism and, therefore, usually a criminal offense and is capable of damaging property (egg whites can degrade certain types of vehicle paint) as well as causing serious eye injury. On Halloween, for example, trick or treaters have been known to throw eggs (and sometimes flour) at property or people from whom they received nothing. Eggs are also often thrown in protests, as they are inexpensive and nonlethal, yet very messy when broken.

Fur

Fur is used in reference to the hair of animals, usually mammals, particularly those with extensive body hair coverage that is generally soft and thick, as opposed to the stiffer bristles on most pigs. The term *pelage* – first known use in English c. 1828 – (French, from Middle French, from *poil* hair, from Old French *peilss*, from Latin *pilus*,) is sometimes used to refer to the body hair of an animal as a complete coat. *Fur* is also used to refer to animal pelts which have been processed into leather with the hair still attached. The words *fur* or *furry* are also used, more casually, to refer to hair-like growths or formations, particularly when the subject being referred to exhibits a dense coat of fine, soft "hairs." If layered, rather than grown as a single coat, it may consist of short down hairs, long guard hairs, and, in some cases, medium awn hairs. Mammals with reduced amounts of fur are often called "naked", as with the naked mole-rat, or "hairless", as with hairless dogs.

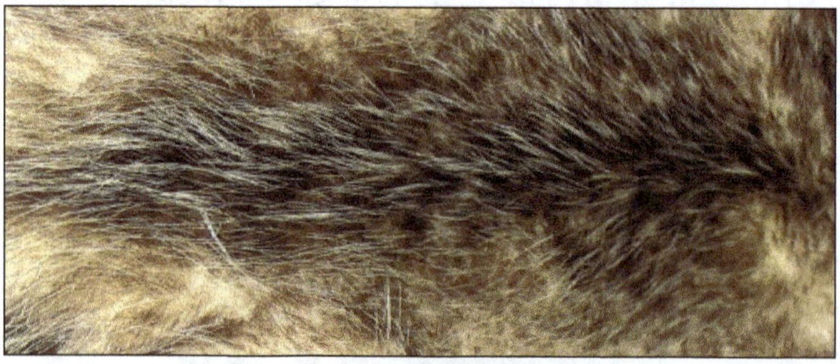

Opossum fur

An animal with commercially valuable fur is known within the fur industry as a furbearer. The use of fur as clothing and/or decoration is considered controversial by some people: most animal welfare advocates object to the trapping and killing of wildlife, and to the confinement and killing of animals on fur farms.

Composition

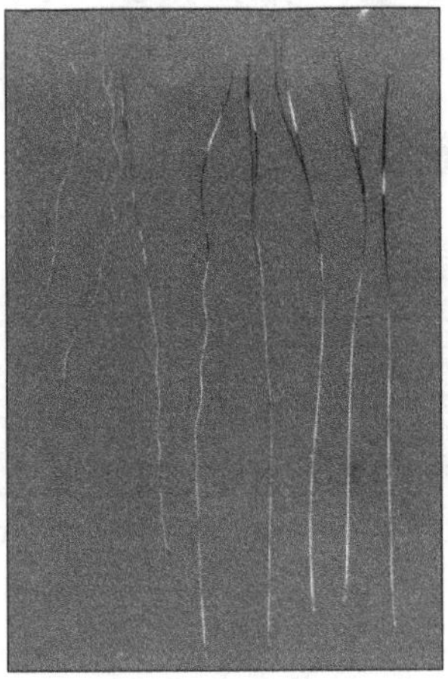

Down, awn and guard hairs of a domestic tabby cat.

Fur usually consists of two main layers:

- Down hair (known also as undercoat or ground hair) — the bottom layer consisting of wool hairs, usually wavy or curly without straight portions or sharp points; down hairs tend to be shorter, flat, curly, and more numerous than the top layer. Its principal function is thermoregulation; it maintains a layer of dry air next to the skin and repels water, thus providing thermal insulation.
- Guard hair — the top layer consisting of longer, generally coarser, nearly straight shafts of hair that protrude through the down hair layer. The distal ends of the guard hairs provide the externally visible layer of the coat of most mammals with well-developed fur. This layer of the coat displays the most marked pigmentation and gloss, including coat patterns adapted to display or camouflage. It is also adapted to shedding water and blocking sunlight, protecting the undercoat and skin from external factors such as rain and ultraviolet radiation. Many animals, such as domestic cats, erect their guard hairs as part of their threat display when agitated.

Mammals with well-developed down and guard hairs also usually have large numbers of awn hairs. These begin their growth much as guard hairs do, but change their mode of growth, usually when less than half the length of the hair has emerged. This portion of the hair is called awn. The rest of the growth is thin and wavy, much like down hair. In many species of mammals, the awn hairs comprise the bulk of the visible coat. The proximal part of the awn hair shares the function of the down hairs, whereas the distal part aids the water-shedding function of the guard hairs, though their thin basal portion prevents their being erected like true guard hairs.

The modern fur arrangement is known to have occurred as far back as docodonts, haramiyidans and eutriconodonts, with *Castorocauda*, *Megaconus* and *Spinolestes* preserving compound follicles with guard hair and underfur.

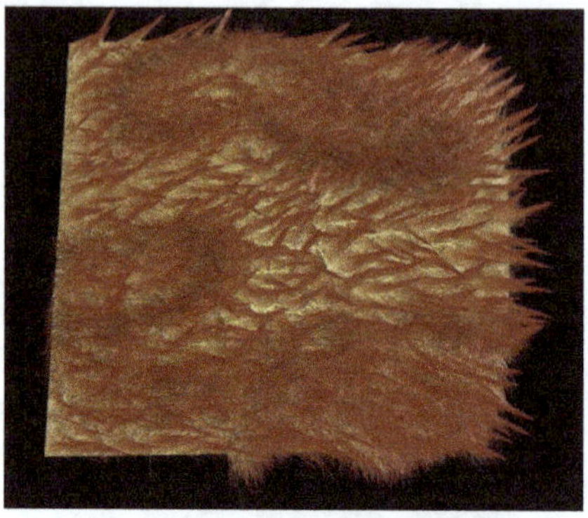

Computer-generated wet fur.

Mammals without Fur

Hair is one of the defining characteristics of mammals, however, several species or breeds have considerably reduced amounts of fur. These are often called "naked" or "hairless".

Natural Selection

Some mammals naturally have reduced amounts of fur. Some semiaquatic or aquatic mammals such as cetaceans, pinnipeds and hippopotamuses have evolved hairlessness, presumably to reduce resistance through water. The naked mole-rat has evolved hairlessness, perhaps as an adaptation to their subterranean life-style. Two of the largest extant mammals, the elephant and the rhinoceros, are largely hairless. The hairless bat is mostly hairless but does have short bristly hairs around its neck, on its front toes, and around the throat sac, along with fine hairs on the head and tail membrane. Most hairless animals cannot go in the sun for long periods of time, or stay in the cold for too long.

Humans are the only primate species that have undergone significant hair loss. The hairlessness of humans compared to related species may be due to loss of functionality in the pseudogene KRTHAP1 (which helps produce keratin) in the human lineage about 240,000 years ago. Mutations in the gene HR can lead to complete hair loss, though this is not typical in humans.

Sheep have not become hairless, however, their pelage is usually referred to as "wool" rather than fur.

Artificial Selection

Humans have artificially selected some domesticated mammalian species to have breeds that are hairless. There are several breeds of hairless cats, perhaps the most commonly known being the Sphynx cat. Similarly, there are several breeds of hairless dogs. Other examples of artificially selected hairless animals include the hairless guinea-pig, nude mouse, and the hairless rat.

Use in Clothing

In clothing, fur is usually leather with the hair retained for its aesthetic and insulating properties. Fur has long served as a source of clothing for hominoids including the Neanderthal. Animal furs used in garments and trim may be dyed bright colors or to mimic exotic animal patterns, or shorn down to imitate the feel of a soft velvet fabric. The term "a fur" is often used to refer to a fur coat, wrap, or shawl.

Carl Ben Eielson, US Pilot and Arctic explorator wearing a seal fur coat.

Usual animal sources for fur clothing and fur trimmed accessories include fox, rabbit, mink, beavers, ermine, otters, sable, seals, coyotes, chinchilla, raccoon, and possum. The import and sale of seal products was banned in the U.S. in 1972 over conservation concerns about Canadian seals. The import and sale is still banned even though the Marine Animal Response Society estimates the harp seal population is thriving at approximately 8 million. The import, export and sales of domesticated cat and dog fur were also banned in the U.S. under the Dog and Cat Protection Act of 2000.

The manufacturing of fur clothing involves obtaining animal pelts where the hair is left on the animal's processed skin. In contrast, making leather involves removing the hair from the hide or pelt and using only the skin. The use of wool involves shearing the animal's fleece from the living animal, so that the wool can be regrown but sheepskin shearling is made by retaining the fleece to the leather and shearing it. Shearling is used for boots, jackets and coats and is probably the most common type of skin worn.

Fur is also used to make felt. A common felt is made from beaver fur and is used in high-end cowboy hats.

Controversy

Most animal rights activists are opposed to the trapping and killing of wildlife, and the confinement and killing of animals on fur farms. According to Humane Society International, over 8 million animals are trapped yearly for fur, while more than 30 million were raised in fur farms.

Red fox furs.

According to Statistics Canada, 2.6 million fur-bearing animals raised on farms were killed in 2010. Another 700,000 were killed for fur by traps.

Wool

Wool is the textile fiber obtained from sheep and certain other animals, including cashmere from goats, mohair from goats, qiviut from muskoxen, angora from rabbits, and other types of wool from camelids.

Wool just before processing.

Sheep before shearing.

Merino sheep

Wool has several qualities that distinguish it from hair or fur: it is crimped, it is elastic, and it grows in staples (clusters).

Characteristics

Wool's scaling and crimp make it easier to spin the fleece by helping the individual fibers attach to each other, so they stay together. Because of the crimp, wool fabrics have greater bulk than other textiles, and they hold air, which causes the fabric to retain heat. Wool has a high specific heat coefficient, so it impedes heat transfer in general. This effect has benefited desert peoples, as Bedouins and Tuaregs use wool clothes for insulation.

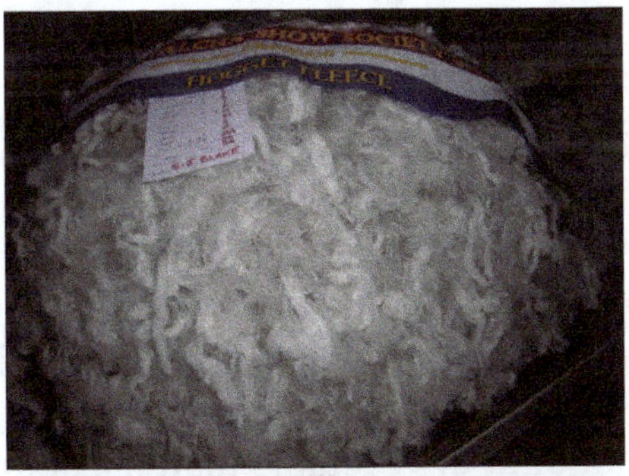

Champion hogget fleece, Walcha Show.

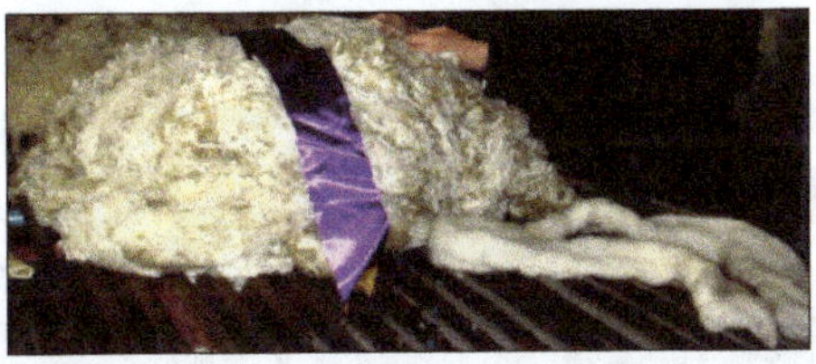

Fleece of fine New Zealand Merino wool and combed wool top on a wool table.

Felting of wool occurs upon hammering or other mechanical agitation as the microscopic barbs on the surface of wool fibers hook together.

The amount of crimp corresponds to the fineness of the wool fibers. A fine wool like Merino may have up to 100 crimps per inch, while the coarser wools like karakul may have as few as one or two. In contrast, hair has little if any scale and no crimp, and little ability to bind into yarn. On sheep, the hair part of the fleece is called kemp. The relative amounts of kemp to wool vary from breed to breed and make some fleeces more desirable for spinning, felting, or carding into batts for quilts or other insulating products, including the famous tweed cloth of Scotland.

Wool fibers readily absorb moisture, but are not hollow. Wool can absorb almost one-third of its own weight in water. Wool absorbs sound like many other fabrics. It is generally a creamy white color, although some breeds of sheep produce natural colors, such as black, brown, silver, and random mixes.

Wool ignites at a higher temperature than cotton and some synthetic fibers. It has a lower rate of flame spread, a lower rate of heat release, a lower heat of combustion, and does not melt or drip; it forms a char which is insulating and self-extinguishing, and it contributes less to toxic gases and smoke than other flooring products when used in carpets. Wool carpets are specified for high safety environments, such as trains and aircraft. Wool is usually specified for garments for firefighters, soldiers, and others in occupations where they are exposed to the likelihood of fire.

Wool is considered by the medical profession to be allergenic.

Processing

Shearing

Sheep shearing is the process by which the woolen fleece of a sheep is cut off. After shearing, the wool is separated into four main categories: fleece (which makes up the vast bulk), broken, bellies, and locks. The quality of fleeces is determined by a technique known as wool classing, whereby a qualified person called a wool classer groups wools of similar gradings together to maximize the return for the farmer or sheep owner. In Australia before being auctioned, all Merino fleece wool is objectively measured for micron, yield (including the amount of vegetable matter), staple length, staple strength, and sometimes color and comfort factor. The sheep is given a dip in antiseptic solution after shearing, so as to cure the wounds caused during shearing.

Fine Merino shearing Lismore, Victoria.

Scouring

Wool straight off a sheep, known as "greasy wool" or "wool in the grease", contains a high level of valuable lanolin, as well as the sheep's dead skin and sweat residue, and generally also contains pesticides and vegetable matter from the animal's environment. Before the wool can be used for commercial purposes, it must be scoured, a process of cleaning the greasy wool. Scouring may be as simple as a bath in warm water or as complicated as an industrial process using detergent and alkali in specialized equipment. In north west England, special potash pits were constructed to produce potash used in the manufacture of a soft soap for scouring locally produced white wool.

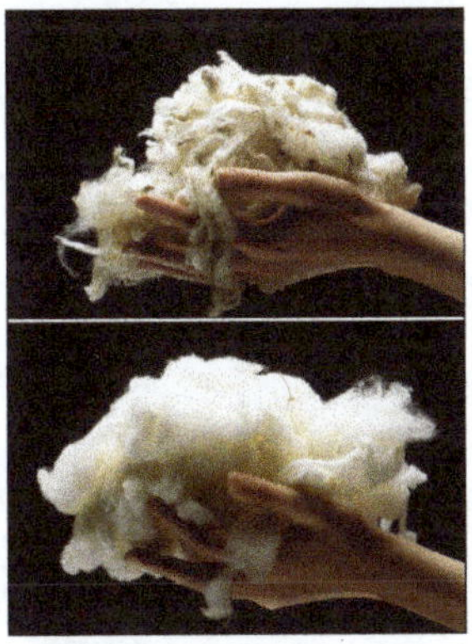

Wool before and after scouring.

In commercial wool, vegetable matter is often removed by chemical carbonization. In less-processed wools, vegetable matter may be removed by hand and some of the lanolin left intact through the use of gentler detergents. This semigrease wool can be worked into yarn and knitted into particularly water-resistant mittens or sweaters, such as those of the Aran Island fishermen. Lanolin removed from wool is widely used in cosmetic products such as hand creams.

Quality

The quality of wool is determined by its fiber diameter, crimp, yield, color, and staple strength. Fiber diameter is the single most important wool characteristic determining quality and price.

Various types and natural colors of wool, and a picture made from wool.

Merino wool is typically 3–5 inches in length and is very fine (between 12 and 24 microns). The finest and most valuable wool comes from Merino hoggets. Wool taken from sheep produced for meat is typically more coarse, and has fibers 1.5 to 6 in (38 to 152 mm) in length. Damage or breaks in the wool can occur if the sheep is stressed while it is growing its fleece, resulting in a thin spot where the fleece is likely to break.

Wool is also separated into grades based on the measurement of the wool's diameter in microns and also its style. These grades may vary depending on the breed or purpose of the wool. For example:

- < 15.5: Ultrafine Merino.
- 15.6 – 18.5: Superfine Merino.
- 18.6 – 20: Fine Merino.
- 20.1 – 23: Medium Merino.
- > 23: Strong Merino.
- Comeback: 21–26 microns, white, 90–180 mm long.
- Fine crossbred: 27–31 microns, Corriedales, etc.
- Medium crossbred: 32–35 microns.
- Downs: 23–34 microns, typically lacks luster and brightness. Examples, Aussiedown, Dorset Horn, Suffolk, etc.
- Coarse crossbred: >36 microns.
- Carpet wools: 35–45 microns.

Any wool finer than 25 microns can be used for garments, while coarser grades are used for outerwear or rugs. The finer the wool, the softer it is, while coarser grades are more durable and less prone to pilling.

The finest Australian and New Zealand Merino wools are known as 1PP, which is the industry benchmark of excellence for Merino wool 16.9 microns and finer. This style represents the top level of fineness, character, color, and style as determined on the basis of a series of parameters in accordance with the original dictates of British wool as applied today by the Australian Wool Exchange (AWEX) Council. Only a few dozen of the millions of bales auctioned every year can be classified and marked 1PP.

In the United States, three classifications of wool are named in the Wool Products Labeling Act of 1939. "Wool" is "the fiber from the fleece of the sheep or lamb or hair of the Angora or Cashmere goat (and may include the so-called specialty fibers from the hair of the camel, alpaca, llama, and vicuna) which has never been reclaimed from any woven or felted wool product". "Virgin wool" and "new wool" are also used to refer to such never used wool. There are two categories of recycled wool (also called reclaimed or shoddy wool). "Reprocessed wool" identifies "wool which has been woven or felted into a wool product and subsequently reduced to a fibrous state without having been used by the ultimate consumer". "Reused wool" refers to such wool that *has* been used by the ultimate consumer.

History

Wild sheep were more hairy than woolly. Although sheep were domesticated some 9,000 to 11,000 years ago, archaeological evidence from statuary found at sites in Iran suggests selection for woolly sheep may have begun around 6000 BC, with the earliest woven wool garments having only been dated to two to three thousand years later. Woolly-sheep were introduced into Europe from the Near East in the early part of the 4th millennium BC. The oldest known European wool textile, ca. 1500 BC, was preserved in a Danish bog. Prior to invention of shears—probably in the Iron Age—the wool was plucked out by hand or by bronze combs. In Roman times, wool, linen, and leather clothed the European population; cotton from India was a curiosity of which only naturalists had heard, and silks, imported along the Silk Road from China, were extravagant luxury goods. Pliny the Elder records in his *Natural History* that the reputation for producing the finest wool was enjoyed by Tarentum, where selective breeding had produced sheep with superior fleeces, but which required special care.

A 1905 illustration of a Tibetan man spinning wool.

In medieval times, as trade connections expanded, the Champagne fairs revolved around the production of wool cloth in small centers such as Provins. The network developed by the annual fairs meant the woolens of Provins might find their way to Naples, Sicily, Cyprus, Majorca, Spain, and even Constantinople. The wool trade developed into serious business, a generator of capital. In the 13th century, the wool trade became the economic engine of the Low Countries and central Italy. By the end of the 14th century, Italy predominated, though Italian production turned to silk in the 16th century. Both industries, based on the export of English raw wool, were rivaled only by the 15th-century sheepwalks of Castile and were a significant source of income to the English crown, which in 1275 had imposed an export tax on wool called the "Great Custom". The importance of wool to the English economy can be seen in the fact that since the 14th century, the presiding officer of the House of Lords has sat on the "Woolsack", a chair stuffed with wool.

Economies of scale were instituted in the Cistercian houses, which had accumulated great tracts of land during the 12th and early 13th centuries, when land prices were low and labor still scarce. Raw wool was baled and shipped from North Sea ports to the textile cities of Flanders, notably Ypres and Ghent, where it was dyed and worked up as cloth. At the time of the Black Death, English textile industries accounted for about 10% of English wool production; the English textile trade grew during the 15th century, to the point where export of wool was discouraged. Over the centuries, various British laws controlled the wool trade or required the use of wool even in burials. The smuggling of wool out of the country, known as owling, was at one time punishable by the cutting off of a hand. After the Restoration, fine English woolens began to compete with silks in the international market, partly aided by the Navigation Acts; in 1699, the English crown forbade its American colonies to trade wool with anyone but England herself.

A great deal of the value of woolen textiles was in the dyeing and finishing of the woven product. In each of the centers of the textile trade, the manufacturing process came to be subdivided into a collection of trades, overseen by an entrepreneur in a system called by the English the "putting-out" system, or "cottage industry", and the *Verlagssystem* by the Germans. In this system of producing wool cloth, until recently perpetuated in the production of Harris tweeds, the entrepreneur provides the raw materials and an advance, the remainder being paid upon delivery of the product. Written contracts bound the artisans to specified terms. Fernand Braudel traces the appearance of the system in the 13th-century economic boom, quoting a document of 1275. The system effectively bypassed the guilds' restrictions.

Before the flowering of the Renaissance, the Medici and other great banking houses of Florence had built their wealth and banking system on their textile industry based on wool, overseen by the Arte della Lana, the wool guild: wool textile interests guided Florentine policies. Francesco Datini, the "merchant of Prato", established in 1383 an *Arte della Lana* for that small Tuscan city. The sheepwalks of Castile shaped the landscape and the fortunes of the *meseta* that lies in the heart of the Iberian peninsula; in the 16th century, a unified Spain allowed export of Merino lambs only with royal permission. The German wool market – based on sheep of Spanish origin – did not overtake British wool until comparatively late. The Industrial Revolution introduced mass production technology into wool and wool cloth manufacturing. Australia's colonial economy was based on sheep raising, and the Australian wool trade eventually overtook that of the Germans by 1845, furnishing wool for Bradford, which developed as the heart of industrialized woolens production.

A World War I-era poster sponsored by the United States Department of Agriculture encouraging children to raise sheep to provide needed war supplies.

Due to decreasing demand with increased use of synthetic fibers, wool production is much less than what it was in the past. The collapse in the price of wool began in late 1966 with a 40% drop; with occasional interruptions, the price has tended down. The result has been sharply reduced production and movement of resources into production of other commodities, in the case of sheep growers, to production of meat.

Superwash wool (or washable wool) technology first appeared in the early 1970s to produce wool that has been specially treated so it is machine washable and may be tumble-dried. This wool is produced using an acid bath that removes the "scales" from the fiber, or by coating the fiber with a polymer that prevents the scales from attaching to each other and causing shrinkage. This process results in a fiber that holds longevity and durability over synthetic materials, while retaining its shape.

In December 2004, a bale of the then world's finest wool, averaging 11.8 microns, sold for AU$3,000 per kilogram at auction in Melbourne, Victoria. This fleece wool tested with an average yield of 74.5%, 68 mm long, and had 40 newtons per kilotex strength. The result was A$279,000 for the bale. The finest bale of wool ever auctioned was sold for a seasonal record of AU$2690 per kilo during June 2008. This bale was produced by the Hillcreston Pinehill Partnership and measured 11.6 microns, 72.1% yield, and had a 43 newtons per kilotex strength measurement. The bale realized $247,480 and was exported to India.

In 2007, a new wool suit was developed and sold in Japan that can be washed in the shower, and which dries off ready to wear within hours with no ironing required. The suit was developed using Australian Merino wool, and it enables woven products made from wool, such as suits, trousers, and skirts, to be cleaned using a domestic shower at home.

In December 2006, the General Assembly of the United Nations proclaimed 2009 to be the International Year of Natural Fibres, so as to raise the profile of wool and other natural fibers.

Production

Global wool production is about 2 million tonnes per year, of which 60% goes into apparel. Australia is a leading producer of wool which is mostly from Merino sheep, but has recently been eclipsed by China in terms of total weight. New Zealand is now (2016) the third-largest producer of wool, and the largest producer of crossbred wool. Breeds such as Lincoln, Romney, Drysdale, and Elliotdale produce coarser fibers, and wool from these sheep is usually used for making carpets.

In the United States, Texas, New Mexico, and Colorado have large commercial sheep flocks and their mainstay is the Rambouillet (or French Merino). Also, a thriving home-flock contingent of small-scale farmers raise small hobby flocks of specialty sheep for the hand-spinning market. These small-scale farmers offer a wide selection of fleece. Global woolclip (total amount of wool shorn) 2004/2005.

1. Australia: 25% of global woolclip (475 million kg greasy, 2004/2005)
2. China: 18%
3. United States: 17%
4. New Zealand: 11%
5. Argentina: 3%
6. Turkey: 2%
7. Iran: 2%
8. United Kingdom: 2%
9. India: 2%
10. Sudan: 2%
11. South Africa: 1%

Organic wool is becoming more and more popular. This wool is very limited in supply and much of it comes from New Zealand and Australia. It is becoming easier to find in clothing and other products, but these products often carry a higher price. Wool is environmentally preferable (as compared to petroleum-based nylon or polypropylene) as a material for carpets, as well, in particular when combined with a natural binding and the use of formaldehyde-free glues.

Animal rights groups have noted issues with the production of wool, such as mulesing.

Marketing

Australia

About 85% of wool sold in Australia is sold by open cry auction. "Sale by sample" is a method in which a mechanical claw takes a sample from each bale in a line or lot of wool. These grab samples are bulked, objectively measured, and a sample of not less than 4 kg is displayed in a box for the buyer to examine. The Australian Wool Exchange conducts sales primarily in Sydney, Melbourne, Newcastle, and Fremantle. About 80 brokers and agents work throughout Australia.

Merino wool samples for sale by auction, Newcastle, New South Wales.

Wool received by Australian brokers and dealers (tonnes/quarter) since 1973.

Wool buyers' room at a wool auction, Newcastle, New South Wales.

About 7% of Australian wool is sold by private treaty on farms or to local wool-handling facilities. This option gives wool growers benefit from reduced transport, warehousing, and selling costs. This method is preferred for small lots or mixed butts to make savings on reclassing and testing.

About 5% of Australian wool is sold over the internet on an electronic offer board. This option gives wool growers the ability to set firm price targets, reoffer passed-in wool, and offer lots to the market quickly and efficiently. This method works well for tested lots, as buyers use these results to make a purchase. About 97% of wool is sold without sample inspection; however, as of December 2009, 59% of wool listed had been passed in from auction. Growers through certain brokers can allocate their wool to a sale and at what price their wool will be reserved.

Sale by tender can achieve considerable cost savings on wool clips large enough to make it worthwhile for potential buyers to submit tenders. Some marketing firms sell wool on a consignment basis, obtaining a fixed percentage as commission.

Forward selling: Some buyers offer a secure price for forward delivery of wool based on estimated measurements or the results of previous clips. Prices are quoted at current market rates and are locked in for the season. Premiums and discounts are added to cover variations in micron, yield, tensile strength, etc., which are confirmed by actual test results when available.

Another method of selling wool includes sales direct to wool mills.

Other Countries

The British Wool Marketing Board operates a central marketing system for UK fleece wool with the aim of achieving the best possible net returns for farmers.

Less than half of New Zealand's wool is sold at auction, while around 45% of farmers sell wool directly to private buyers and end-users.

United States sheep producers market wool with private or cooperative wool warehouses, but wool pools are common in many states. In some cases, wool is pooled in a local market area, but sold through a wool warehouse. Wool offered with objective measurement test results is preferred. Imported apparel wool and carpet wool goes directly to central markets, where it is handled by the large merchants and manufacturers.

Yarn

Virgin wool is spun for the first time.

Worsted wool yarn, the first step in the manufacture of most wool clothing.

Shoddy or recycled wool is made by cutting or tearing apart existing wool fabric and respinning the resulting fibers. As this process makes the wool fibers shorter, the remanufactured fabric is inferior to the original. The recycled wool may be mixed with raw wool, wool noil, or another fiber such as cotton to increase the average fiber length. Such yarns are typically used as weft yarns with a cotton warp. This process was invented in the Heavy Woollen District of West Yorkshire and created a microeconomy in this area for many years.

Rag is a sturdy wool fiber made into yarn and used in many rugged applications such as gloves.

Worsted is a strong, long-staple, combed wool yarn with a hard surface.

Woolen is a soft, short-staple, carded wool yarn typically used for knitting. In traditional weaving, woolen weft yarn (for softness and warmth) is frequently combined with a worsted warp yarn for strength on the loom.

Uses

In addition to clothing, wool has been used for blankets, horse rugs, saddle cloths, carpeting, felt, wool insulation and upholstery. Wool felt covers piano hammers, and it is used to absorb odors and noise in heavy machinery and stereo speakers. Ancient Greeks lined their helmets with felt, and Roman legionnaires used breastplates made of wool felt.

Woolen garments in the wool samples area of a wool store, Newcastle, New South Wales.

Wool has also been traditionally used to cover cloth diapers. Wool fiber exteriors are hydrophobic (repel water) and the interior of the wool fiber is hygroscopic (attracts water); this makes a wool garment suitable cover for a wet diaper by inhibiting wicking, so outer garments remain dry. Wool felted and treated with lanolin is water resistant, air permeable, and slightly antibacterial, so it resists the buildup of odor. Some modern cloth diapers use felted wool fabric for covers, and there are several modern commercial knitting patterns for wool diaper covers.

Initial studies of woolen underwear have found it prevented heat and sweat rashes because it more readily absorbs the moisture than other fibers.

Merino wool has been used in baby sleep products such as swaddle baby wrap blankets and infant sleeping bags.

As an animal protein, it can be used as a soil fertilizer, being a slow-release source of nitrogen.

Researchers at the Royal Melbourne Institute of Technology school of fashion and textiles have discovered a blend of wool and kevlar, the synthetic fiber widely used in body armor, was lighter, cheaper and worked better in damp conditions than kevlar alone. Kevlar, when used alone, loses about 20% of its effectiveness when wet, so required an expensive waterproofing process. Wool increased friction in a vest with 28–30 layers of fabric, to provide the same level of bullet resistance as 36 layers of Kevlar alone.

Carbon Footprint

The carbon footprint of woolen products is difficult to determine and varies considerably according to factors ranging from the conditions in which sheep are bred to the processes used to manufacture the products.

The average greenhouse gas footprint of wool in manufacturing carpets estimated at 5.48 kg CO_2 equivalent per kg, when produced in Europe.

The carbon footprint of a 264.85g woolly jumper made from New Zealand merino wool averages about 1.667 kg CO_2 equivalent at the point of purchase by a consumer, implying a carbon footprint of around 6 kg per kg of finished woolen product.

Events

A buyer of Merino wool, Ermenegildo Zegna, has offered awards for Australian wool producers. In 1963, the first Ermenegildo Zegna Perpetual Trophy was presented in Tasmania for growers of "Superfine skirted Merino fleece". In 1980, a national award, the Ermenegildo Zegna Trophy for Extrafine Wool Production, was launched. In 2004, this award became known as the Ermenegildo Zegna Unprotected Wool Trophy. In 1998, an Ermenegildo Zegna Protected Wool Trophy was launched for fleece from sheep coated for around nine months of the year.

Andean woman sorting wool as part of the theme park Los Aleros in Mérida, Venezuela.

In 2002, the Ermenegildo Zegna Vellus Aureum Trophy was launched for wool that is 13.9 microns or finer. Wool from Australia, New Zealand, Argentina, and South Africa may enter, and a winner is named from each country. In April 2008, New Zealand won the Ermenegildo Zegna Vellus Aureum Trophy for the first time with a fleece that measured 10.8 microns. This contest awards the winning fleece weight with the same weight in gold as a prize, hence the name.

In 2010, an ultrafine, 10-micron fleece, from Windradeen, near Pyramul, New South Wales, won the Ermenegildo Zegna Vellus Aureum International Trophy.

Since 2000, Loro Piana has awarded a cup for the world's finest bale of wool that produces just enough fabric for 50 tailor-made suits. The prize is awarded to an Australian or New Zealand wool grower who produces the year's finest bale.

The New England Merino Field days which display local studs, wool, and sheep are held during January, in even numbered years around the Walcha, New South Wales district. The Annual Wool Fashion Awards, which showcase the use of Merino wool by fashion designers, are hosted by the city of Armidale, New South Wales, in March each year. This event encourages young and established fashion designers to display their talents. During each May, Armidale hosts the annual New England Wool Expo to display wool fashions, handicrafts, demonstrations, shearing competitions, yard dog trials, and more.

In July, the annual Australian Sheep and Wool Show is held in Bendigo, Victoria. This is the largest sheep and wool show in the world, with goats and alpacas, as well as woolcraft competitions and displays, fleece competitions, sheepdog trials, shearing, and wool handling. The largest competition in the world for objectively measured fleeces is the Australian Fleece Competition, which is held annually at Bendigo. In 2008, 475 entries came from all states of Australia, with first and second prizes going to the Northern Tablelands, New South Wales fleeces.

Honey

Honey is a sweet food made by bees foraging nectar from flowers. The variety produced by honey bees (the genus *Apis*) is the one most commonly referred to, as it is the type of honey collected by most beekeepers and consumed by people. Honey is also produced by bumblebees, stingless bees, and other hymenopteran insects such as honey wasps, though the quantity is generally lower and they have slightly different properties compared to honey from the genus *Apis*. Honey bees convert nectar into honey by a process of regurgitation and evaporation: they store it as a primary food source in wax honeycombs inside the beehive.

Honey gets its sweetness from the monosaccharides fructose and glucose, and has about the same relative sweetness as granulated sugar. It has attractive chemical properties for baking and a distinctive flavor that leads some people to prefer it to sugar and other sweeteners. Most microorganisms do not grow in honey so sealed honey does not spoil, even after thousands of years. However, honey sometimes contains dormant endospores of the bacterium *Clostridium botulinum*, which can be dangerous to babies, as it may result in botulism.

A jar of honey with a honey dipper and an American biscuit.

People who have a weakened immune system should not eat honey because of the risk of bacterial or fungal infection. There is some evidence that honey may be effective in treating diseases and other medical conditions such as wounds and burns. However, the evidence is overall not conclusive. Providing 64 calories in a typical serving of one tablespoon (15 mL), honey contains no significant essential nutrient content. Honey is generally safe but there are various, potential adverse-effects or interactions it may have in combination with excessive consumption, existing disease conditions, or drugs.

Honey in honeycomb.

Honey use and production has a long and varied history. Honey collection is an ancient activity. Humans apparently began foraging for honey at least 8,000 years ago, as evidenced by a cave painting in Valencia, Spain.

Formation

Honey is produced by bees from nectar collection which serves the dual purpose to support metabolism of muscle activity during foraging and for long-term food storage as honey. During foraging, bees access part of the nectar collected to support metabolic activity of flight muscles, with the majority of collected nectar destined for regurgitation, digestion and storage as honey. In cold weather or when other food sources are scarce, adult and larval bees use stored honey as food.

A honey bee on calyx of goldenrod

By contriving for bee swarms to nest in artificial hives, people have been able to semi-domesticate the insects and harvest excess honey. In the hive or in a wild nest, the three types of bees are:

- A single female queen bee.
- A seasonally variable number of male drone bees to fertilize new queens.
- 20,000 to 40,000 female worker bees.

Leaving the hive, foraging bees collect sugar-rich flower nectar and return to the hive where they use their "honey stomachs" to ingest and regurgitate the nectar repeatedly until it is partially digested. Bee digestive enzymes - invertase, amylase and diastase - and gastric acid hydrolyze sucrose to a mixture of glucose and fructose. The bees work together as a group with the regurgitation and digestion for as long as 20 minutes until the product reaches storage quality. It is then placed in honeycomb cells left unsealed while still high in water content (about 20%) and natural yeasts, which, unchecked, would cause the sugars in the newly formed honey to ferment. The process continues as hive bees flutter their wings constantly to circulate air and evaporate water from the honey to a content of about 18%, raising the sugar concentration and preventing fermentation. The bees then cap the cells with wax to seal them. As removed from the hive by a beekeeper, honey has a long shelf life and will not ferment if properly sealed.

Another source of honey is from a number of wasp species, such as the wasps *Brachygastra lecheguana* and *Brachygastra mellifica*, which are found in South and Central America. These species are known to feed on nectar and produce honey.

Some wasps, such as the *Polistes versicolor*, even consume honey themselves, switching from feeding on pollen in the middle of their lifecycles to feeding on honey, which can better provide for their energy needs.

Extraction from a honeycomb.

Filtering from a honeycomb.

Pouring raw honey.

Collection

Honey is collected from wild bee colonies, or from domesticated beehives. Wild bee nests are sometimes located by following a honeyguide bird. The bees may first be pacified by using smoke from a bee smoker. The smoke triggers a feeding instinct (an attempt to save the resources of the hive from a possible fire), making them less aggressive and the smoke obscures the pheromones the bees use to communicate.

The honeycomb is removed from the hive and the honey may be extracted from that, either by crushing or by using a honey extractor. The honey is then usually filtered to remove beeswax and other debris.

Before the invention of removable frames, bee colonies were often sacrificed in order to conduct the harvest. The harvester would take all the available honey and replace the entire colony the next spring. Since the invention of removable frames, the principles of husbandry lead most beekeepers to ensure that their bees will have enough stores to survive the winter, either by leaving some honey in the beehive or by providing the colony with a honey substitute such as sugar water or crystalline sugar (often in the form of a "candyboard"). The amount of food necessary to survive the winter depends on the variety of bees and on the length and severity of local winters.

A wide range of species other than humans are attracted to wild or domestic sources of honey.

Production

Top five honey producing countries (millions of tonnes)		
Rank	**Country**	**2013**
1	China	0.47
2	Turkey	0.09
3	Argentina	0.08
4	Ukraine	0.07
5	Russia	0.07
--	**World**	**1.7**
Source: UN Food & Agriculture Organization, FAOSTAT		

In 2013, 1.7 million tonnes of honey were produced worldwide, with China accounting for 28% of the world total (table). The next four largest producers – Turkey, Argentina, Ukraine and Russia – accounted collectively for less than 20% of the world total (table).

Modern Uses

Food

Over its history as a food, the main uses of honey are in cooking, baking, desserts, such as *mel i mató*, as a spread on bread, and as an addition to various beverages, such as tea, and as a sweetener in some commercial beverages. Honey barbecue and honey mustard are other common flavors used in sauces.

Fermentation

Honey is the main ingredient in the alcoholic beverage mead, which is also known as "honey wine" or "honey beer". Historically, the ferment for mead was honey's naturally occurring yeast. Honey is also used as an adjunct in some beers.

Honey wine, or mead, is typically (modern era) made with a honey and water mixture with yeast added for fermentation. Primary fermentation usually takes 28–56 days, after which the must needs to be racked into a secondary fermentation vessel and left to sit about 35–40 more days. If done properly, fermentation will be finished by this point (though if a sparkling mead is desired, fermentation can be restarted after bottling by the addition of a small amount of sugar), but most meads require aging for 6–9 months or more in order to be palatable.

Physical and Chemical Properties

The physical properties of honey vary, depending on water content, the type of flora used to produce it (pasturage), temperature, and the proportion of the specific sugars it contains. Fresh honey is a supersaturated liquid, containing more sugar than the water can typically dissolve at ambient temperatures. At room temperature, honey is a supercooled liquid, in which the glucose will precipitate into solid granules. This forms a semisolid solution of precipitated glucose crystals in a solution of fructose and other ingredients.

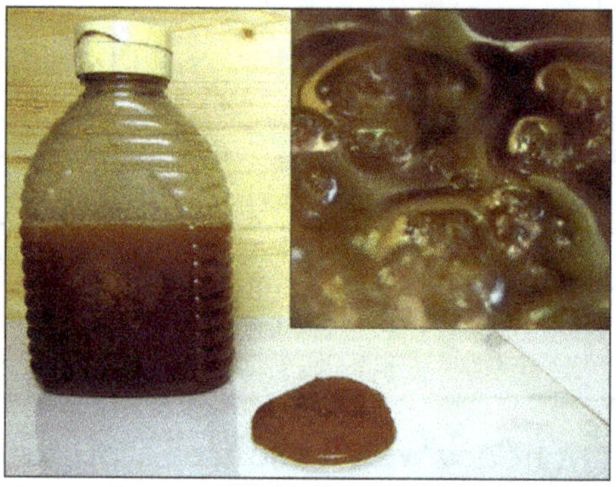

Crystallized honey. The inset shows a close-up of the honey, showing the individual glucose grains in the fructose mixture.

Phase Transitions

The melting point of crystallized honey is between 40 and 50 °C (104 and 122 °F), depending on its composition. Below this temperature, honey can be either in a metastable state, meaning that it will not crystallize until a seed crystal is added, or, more often, it is in a "labile" state, being saturated with enough sugars to crystallize spontaneously. The rate of crystallization is affected by many factors, but the primary factor is the ratio of the main sugars: fructose to glucose. Honeys that are supersaturated with a very high percentage of glucose, such as brassica honey, will crystallize almost immediately after harvesting, while honeys with a low percentage of glucose, such as chestnut or tupelo honey, do not crystallize. Some types of honey may produce very large but few crystals, while others will produce many small crystals.

Crystallization is also affected by water content, because a high percentage of water will inhibit crystallization, as will a high dextrin content. Temperature also affects the rate of crystallization, with the fastest growth occurring between 13 and 17 °C (55 and 63 °F).

Crystal nuclei (seeds) tend to form more readily if the honey is disturbed, by stirring, shaking or agitating, rather than if left at rest. However, the nucleation of microscopic seed-crystals is greatest between 5 and 8 °C (41 and 46 °F). Therefore, larger but fewer crystals tend to form at higher temperatures, while smaller but more-numerous crystals usually form at lower temperatures. Below 5 °C, the honey will not crystallize and, thus, the original texture and flavor can be preserved indefinitely.

Since honey normally exists below its melting point, it is a supercooled liquid. At very low temperatures, honey will not freeze solid. Instead, as the temperatures become lower, the viscosity of honey increases. Like most viscous liquids, the honey will become thick and sluggish with decreasing temperature. At −20 °C (−4 °F), honey may appear or even feel solid, but it will continue to flow at very low rates. Honey has a glass transition between −42 and −51 °C (−44 and −60 °F). Below this temperature, honey enters a glassy state and will become an amorphous solid (noncrystalline).

Viscosity

The viscosity of honey is affected greatly by both temperature and water content. The higher the water percentage, the easier honey flows. Above its melting point, however, water has little effect on viscosity. Aside from water content, the composition of honey also has little effect on viscosity, with the exception of a few types. At 25 °C (77 °F), honey with 14% water content generally has a viscosity around 400 poise, while a honey containing 20% water has a viscosity around 20 poise. Viscosity increase due to temperature occurs very slowly at first. A honey containing 16% water, at 70 °C (158 °F), will have a viscosity around 2 poise, while at 30 °C (86 °F), the viscosity is around 70 poise. As cooling progresses, honey becomes more viscous at an increasingly rapid rate, reaching 600 poise around 14 °C (57 °F). However, while honey is very viscous, it has rather low surface tension.

A few types of honey have unusual viscous properties. Honeys from heather or manuka display thixotropic properties. These types of honey enter a gel-like state when motionless, but then liquify when stirred.

Electrical and Optical Properties

Because honey contains electrolytes, in the form of acids and minerals, it exhibits varying degrees of electrical conductivity. Measurements of the electrical conductivity are used to determine the quality of honey in terms of ash content.

The effect honey has on light is useful for determining the type and quality. Variations in the water content alter the refractive index of honey. Water content can easily be measured with a refractometer. Typically, the refractive index for honey will range from 1.504 at 13% water content to 1.474 at 25%. Honey also has an effect on polarized light, in that it will rotate the polarization plane. The fructose will give a negative rotation, while the glucose will give a positive one. The overall rotation can be used to measure the ratio of the mixture. Honey may vary in color between pale yellow and dark brown, but other bright colors may occasionally be found, depending on the source of the sugar harvested by the bees.

Hygroscopy and Fermentation

Honey has the ability to absorb moisture directly from the air, a phenomenon called hygroscopy. The amount of water the honey will absorb is dependent on the relative humidity of the air. Because honey contains yeast, this hygroscopic nature requires that honey be stored in sealed containers to prevent fermentation, which usually begins if the honey's water content rises much above 25%. Honey will tend to absorb more water in this manner than the individual sugars would allow on their own, which may be due to other ingredients it contains.

Fermentation of honey will usually occur after crystallization because, without the glucose, the liquid portion of the honey primarily consists of a concentrated mixture of the fructose, acids, and water, providing the yeast with enough of an increase in the water percentage for growth. Honey that is to be stored at room temperature for long periods of time is often pasteurized, to kill any yeast, by heating it above 70 °C (158 °F).

Thermal Characteristics

Like all sugar compounds, honey will caramelize if heated sufficiently, becoming darker in color, and eventually burn. However, honey contains fructose, which caramelizes at lower temperatures than the glucose. The temperature at which caramelization begins varies, depending on the composition, but is typically between 70 and 110 °C (158 and 230 °F). Honey also contains acids, which act as catalysts, decreasing the caramelization temperature even more. Of these acids, the amino acids, which occur in very small amounts, play an important role in the darkening of honey. The amino acids form darkened compounds called melanoidins, during a Maillard reaction. The Maillard reaction will occur slowly at room temperature, taking from a few to several months to show visible darkening, but will speed-up dramatically with increasing temperatures. However, the reaction can also be slowed by storing the honey at colder temperatures.

Unlike many other liquids, honey has very poor thermal conductivity, taking a long time to reach thermal equilibrium. Melting crystallized honey can easily result in localized caramelization if the heat source is too hot, or if it is not evenly distributed. However, honey will take substantially longer to liquify when just above the melting point than it will at elevated temperatures. Melting 20 kilograms of crystallized honey, at 40 °C (104 °F), can take up to 24 hours, while 50 kilograms may take twice as long. These times can be cut nearly in half by heating at 50 °C (122 °F). However, many of the minor substances in honey can be affected greatly by heating, changing the flavor, aroma, or other properties, so heating is usually done at the lowest temperature possible for the shortest amount of time.

References

- Schor, Juliet; Taylor, Betsy (editors). Sustainable Planet: Roadmaps for the Twenty-First Century. Beacon Press, 2003. ISBN 0-8070-0455-3
- Van den Bossche, Peter. The Law and Policy of the bosanac Trade Organization: Text, Cases and Materials. UK: Cambridge University Press, 2005. ISBN 0-521-82290-4
- Serope Kalpakjian, Steven R Schmid. "Manufacturing Engineering and Technology". International edition. 4th Ed. Prentice Hall, Inc. 2001. ISBN 0-13-017440-8
- Magdoff, Fred; Foster, John Bellamy; and Buttel, Frederick H. Hungry for Profit: The Agribusiness Threat to Farmers, Food, and the Environment. September 2000. ISBN 1-58367-016-5
- McGee, Harold. On Food and Cooking: The Science and Lore of the Kitchen. New York: Simon & Schuster, 2004. ISBN 0-684-80001-2
- Braaten, Ann W. (2005). "Wool". In Steele, Valerie. Encyclopedia of Clothing and Fashion. 3. Thomson Gale. pp. 441–443. ISBN 0-684-31394-4
- Parekh, Sarad R. The Gmo Handbook: Genetically Modified Animals, Microbes, and Plants in Biotechnology. Humana Press,2004. ISBN 1-58829-307-6

Permissions

All chapters in this book are published with permission under the Creative Commons Attribution Share Alike License or equivalent. Every chapter published in this book has been scrutinized by our experts. Their significance has been extensively debated. The topics covered herein carry significant information for a comprehensive understanding. They may even be implemented as practical applications or may be referred to as a beginning point for further studies.

We would like to thank the editorial team for lending their expertise to make the book truly unique. They have played a crucial role in the development of this book. Without their invaluable contributions this book wouldn't have been possible. They have made vital efforts to compile up to date information on the varied aspects of this subject to make this book a valuable addition to the collection of many professionals and students.

This book was conceptualized with the vision of imparting up-to-date and integrated information in this field. To ensure the same, a matchless editorial board was set up. Every individual on the board went through rigorous rounds of assessment to prove their worth. After which they invested a large part of their time researching and compiling the most relevant data for our readers.

The editorial board has been involved in producing this book since its inception. They have spent rigorous hours researching and exploring the diverse topics which have resulted in the successful publishing of this book. They have passed on their knowledge of decades through this book. To expedite this challenging task, the publisher supported the team at every step. A small team of assistant editors was also appointed to further simplify the editing procedure and attain best results for the readers.

Apart from the editorial board, the designing team has also invested a significant amount of their time in understanding the subject and creating the most relevant covers. They scrutinized every image to scout for the most suitable representation of the subject and create an appropriate cover for the book.

The publishing team has been an ardent support to the editorial, designing and production team. Their endless efforts to recruit the best for this project, has resulted in the accomplishment of this book. They are a veteran in the field of academics and their pool of knowledge is as vast as their experience in printing. Their expertise and guidance has proved useful at every step. Their uncompromising quality standards have made this book an exceptional effort. Their encouragement from time to time has been an inspiration for everyone.

The publisher and the editorial board hope that this book will prove to be a valuable piece of knowledge for students, practitioners and scholars across the globe.

Index

www.ingramcontent.com/pod-product-compliance
Lightning Source LLC
LaVergne TN
LVHW080623210926
848485LV00003B/207